Nonlinear Phenomena in the Radiation from Plasmas

Online at: https://doi.org/10.1088/978-0-7503-5552-0

Nonlinear Phenomena in the Radiation from Plasmas

Eugene Oks
Physics Department, Auburn University, Auburn, AL, USA

IOP Publishing, Bristol, UK

ISBN 978-0-7503-5552-0 (ebook)
ISBN 978-0-7503-5550-6 (print)
ISBN 978-0-7503-5553-7 (myPrint)
ISBN 978-0-7503-5551-3 (mobi)

DOI 10.1088/978-0-7503-5552-0

Version: 20230701

IOP ebooks

British Library Cataloguing-in-Publication Data: A catalogue record for this book is available from the British Library.

Published by IOP Publishing, wholly owned by The Institute of Physics, London

IOP Publishing, No.2 The Distillery, Glassfields, Avon Street, Bristol, BS2 0GR, UK

US Office: IOP Publishing, Inc., 190 North Independence Mall West, Suite 601, Philadelphia, PA 19106, USA

To Irisha with love

Contents

About the author

Eugene Oks

Eugene Oks received his PhD degree from the Moscow Institute of Physics and Technology and later received the highest degree of Doctor of Sciences from the Institute of General Physics of the Academy of Sciences of the USSR by the decision of the Scientific Council led by the Nobel Prize winner, academician A M Prokhorov. According to the Statute of the Doctor of Sciences degree, this highest degree is awarded only to the most outstanding PhD scientists who have founded a new research field of great interest. Oks worked in Moscow (USSR) as the head of a research unit at the Center for Studying Surfaces and Vacuum; he then worked at the Ruhr University in Bochum (Germany) as an invited professor, and for the last 30 plus years, he has worked at the Physics Department of Auburn University (USA) in the position of Professor. He has conducted research in five areas: atomic and molecular physics, astrophysics, plasma physics, laser physics, and nonlinear dynamics. He has founded/co-founded and developed new research fields, such as intra-Stark spectroscopy (a new class of nonlinear optical phenomena in plasmas), masing without inversion (advanced schemes for generating/amplifying coherent microwave radiation), and quantum chaos (nonlinear dynamics in the microscopic world). He has also developed a large number of advanced spectroscopic methods for diagnosing various laboratory and astrophysical plasmas—methods that were then used and are still used by many experimental groups around the world. He recently revealed that there are two flavors of hydrogen atoms, as proven by the analysis of atomic experiments; there is also a possible astrophysical proof from observations of the anomalous absorption of the 21 cm radio line in the early Universe and from the observed over-smooth distribution of dark matter in the Universe. He has shown that dark matter—or at least a part of it—can be attributed to the second flavor of hydrogen atom. He has published about 600 papers and 12 books, including the books 'Plasma Spectroscopy: The Influence of Microwave and Laser Fields,' 'Stark Broadening of Hydrogen and Hydrogenlike Spectral Lines in Plasmas: The Physical Insight,' 'Breaking Paradigms in Atomic and Molecular Physics,' 'Diagnostics of Laboratory and Astrophysical Plasmas Using Spectral Lineshapes of One-, Two, and Three-Electron Systems,' 'Unexpected Similarities of the Universe with Atomic and Molecular Systems: What a Beautiful World,' 'Analytical Advances in Quantum and Celestial Mechanics: Separating Rapid and Slow Subsystems,' 'Advances in X-Ray Spectroscopy of Laser Plasmas,' 'Simple Atomic and Molecular Systems: New Results and Applications,' 'Advances in the Physics of Rydberg Atoms and Molecules,' 'The Second Flavor of Hydrogen Atoms—the Leading Candidate for Dark Matter: Theoretical Discovery and the Proofs from Experiments and Astrophysical Observations,' and 'Recent Advances in Dark Matter, Dark Energy, Exoplanets, Flare Stars, White Dwarfs, and More.' He is the editor-in-chief of the physical sciences section of the journal 'Foundations' and the editor-in-chief of the journal 'International Review of Atomic and Molecular Physics' and of the journal 'Physics International.' He is a member of the editorial boards of six other journals: 'Symmetry,' 'American Journal of Astronomy and Astrophysics,' 'Dynamics,' 'Open Physics,' 'Open Journal of Microphysics,' and 'Current Physics.' He is a member of the reviewers board of the journal 'Atoms.' He is also a member of the International Program Committees of the biannual series of conferences known as 'Spectral Line Shapes.'

IOP Publishing

Nonlinear Phenomena in the Radiation from Plasmas

Eugene Oks

Chapter 1

Introduction

The most informative way to study laboratory and astrophysical plasmas is by analyzing their radiation. Nonlinear phenomena are encountered almost everywhere in nature and in laboratory experiments. The overturning of a wave coming into a beach is an example of nonlinearity in nature that most people have seen: the wave becomes nonlinear. A ubiquitous example of man-made nonlinearity can be found in electric circuits: as the current heats the wire, its resistance changes, so that the linear Ohm's law ceases to be valid.

The radiation from plasmas is no exception. The nonlinear phenomena in the radiation from plasmas are important both fundamentally as well as practically and are the subject of this book. The book is structured as follows.

Chapter 2 presents an overview of the studies of nonlinear effects in the radiation of satellites of hydrogenic spectral lines from plasmas and their applications. The satellites are caused by various types of quasimonochromatic electric field. These fields can be either intrinsic to plasmas (such as Langmuir waves) or extrinsic (such as a laser field). The effects under consideration are nonlinear with respect to the energy density $E_0^2/(16\pi)$ of a quasimonochromatic electric field of amplitude E_0. This chapter covers nonlinear effects in the following situations: (1) satellites under a one-dimensional one-mode monochromatic electric field in the non-opaque case; (2) satellites under a one-dimensional one-mode monochromatic electric field in the opaque case; (3) satellites under a one-dimensional two-mode monochromatic electric field; (4) satellites under a one-dimensional multimode monochromatic electric field; (5) satellites under a two-dimensional multimode electric field; (6) satellites in an elliptically polarized electric field; (7) nonlinear spectral effects in the case of Langmuir solitons.

Chapter 3 presents nonlinear effects in the radiation of hydrogenic spectral lines under the broadband electric field of Langmuir turbulence. The Langmuir waves, which are quasimonochromatic in plasmas of relatively low electron densities N_e, can be broadband waves (turbulent) at high electron densities, such as $N_e \gg 10^{18}$ cm^{-3}.

The broadening of the power spectrum of the Langmuir waves at high electron densities is mainly caused by electron collisions. We consider two possible shapes of the power spectrum of the stochastic electric field (such as broadband Langmuir turbulence): the Lorentzian shape and the Gaussian shape. We analyze effects that are nonlinear with respect to the energy density of the stochastic electric field.

Chapter 4 presents a study of various nonlinear effects in the radiation of satellites of non-hydrogenic spectral lines from plasmas and their applications. This kind of satellite can occur in the spectral lines of helium and lithium atoms as well as in the spectral lines of He-like and Li-like ions. We cover the early theoretical results obtained for the situation in which the quasimonochromatic electric field (causing the satellites) is relatively weak, so that the effects can be described in the frame of the perturbation theory. We then cover theoretical and experimental results obtained for stronger fields: the adiabatic theory of satellites and its applications, as well as the case in which the frequency of the quasimonochromatic electric field is much greater than the corresponding atomic frequencies.

Chapter 5 presents the simultaneous interaction between the radiating atom or ion and the quasimonochromatic electric field and with the quasistatic part of the plasma electric field, which led to the discovery of a new sub-area of plasma spectroscopy: *intra-Stark spectroscopy*. It deals with Langmuir-wave-induced structures (bump-dip-bump structures) in certain locations of the profiles of hydrogenic spectral lines. These structures (called for brevity Langmuir 'dips'), consisting of a local intensity minimum surrounded by two 'bumps' (peaks)—are *emergent phenomena* that spring from *multifrequency nonlinear dynamic resonances.* The analytical predictions of the emergent phenomenon of the Langmuir-wave-induced 'dips' have been confirmed by a large number of experiments by various experimental groups working with different plasma machines as well as by astrophysical observations. In these experiments and observations, which span about ten orders of magnitude of electron density, the highly localized Langmuir-wave-induced structures were reliably detected, identified, and used for plasma diagnostics. In particular, this included high-precision, benchmark experiments at the gas-liner pinch, where plasma parameters were measured using coherent Thomson scattering independently of measurements of the line profiles.

Chapter 6 presents nonlinear effects in the radiation of quasistatically Stark-broadened spectral lines from plasmas under a high-frequency laser field $\mathbf{E}_0 \cos \omega t$. The high-frequency laser field partially or completely suppresses the components of the quasistatic field $\mathbf{F}$ perpendicular to $\mathbf{E}_0$. This leads to a highly nonlinear dependence of the intensities of the spectral line components and the shift of spectral line components on the scaled energy density of the high-frequency laser field.

Chapter 7 presents nonlinear effects in the radiation of dynamically Stark-broadened hydrogenic spectral lines by plasma electrons. In analytical theories of the Stark broadening of spectral lines by the electron microfield in plasmas, the effects of the electrons were considered in the dynamic regime. Specifically, Stark broadening was treated as resulting from a sequence of *binary collisions* with the perturbing electrons (some collisions were not completed). Due to the binary nature of this framework, it might be expected that the resulting Stark width would be linear with respect to the plasma electron density N_e. However, in reality, the Stark width turned out to be nonlinear with respect to N_e: either weakly nonlinear, as in the early theories, or strongly nonlinear, as in the more sophisticated theories. This chapter covers the following: (1) early theoretical results; (2) the refinement of the conventional theory for hydrogen spectral lines; (3) the refinement of the conventional theory for hydrogenlike spectral lines; (4) the generalized theory; (5) reduction of Stark broadening due to the acceleration of electrons by the ion field in plasmas.

Chapter 8 presents nonlinear effects in dynamical Stark broadening by plasma ions. We demonstrate that the ion dynamical Stark width of the lines without the central Stark components (such as the Ly-beta, Lyman-delta, Balmer beta, or Balmer-delta lines) depends on the ion density N_i in a significantly nonlinear way. At relatively small ion densities, the dependence is quasi-linear. However, at relatively large ion densities, the dependence becomes significantly nonlinear: proportional to $N_i^{1/2}$. We also show that the ion dynamical Stark width of these spectral lines has the following dependence on the ion temperature T. At relatively large temperatures, the ion dynamical Stark width is proportional to $(1/T)^{1/2}$. At relatively small temperatures, the ion dynamical Stark width is proportional to T^4.

Some additional topics are covered in the appendices. Appendix A exhibits the effects of various electric fields on the energy levels 5D, 5F, and 5G of helium. Appendix B provides practical examples of the use of spectral line radiation to diagnose oscillatory electric fields in laser-produced plasmas. Appendix C is devoted to the validity of using the analytical method based on the separation of rapid and slow subsystems for a more accurate analysis of the Stark broadening of hydrogen-like spectral lines by plasma electrons. Appendix D provides a brief outline of the impact approximation in the conventional theory of the Stark broadening of spectral lines in plasmas.

Major publications related to the topic of this book are listed below.

Further reading

[1] Blochinzew D I 1933 *Phys. Z. Sow. Union* **4** 501

[2] Oks E 2021 *Plasma* **4** 65

[3] Oks E 1995 *Plasma Spectroscopy: The Influence of Microwave and Laser Fields* (Berlin: Springer)

[4] Lifshitz E V 1968 *Sov. Phys. JETP* **26** 570 http://jetp.ras.ru/cgi-bin/e/index/e/26/3/p570?a=list

[5] Zel'dovich J B 1967 *Sov. Phys. JETP* **24** 1006 http://jetp.ras.ru/cgi-bin/e/index/e/24/5/p1006?a=list

[6] Ritus V I 1967 *Sov. Phys. JETP* **24** 1041 http://jetp.ras.ru/cgi-bin/e/index/e/24/5/p1041?a=list

[7] Oks E 2021 *Eur. Phys. J.* D **75** 185

[8] Bernstein I B, Greene J M and Kruskal M D 1957 *Phys. Rev.* **108** 546

[9] Oks E 2017 *Diagnostics of Laboratory and Astrophysical Plasmas Using Spectral Lineshapes of One-, Two-, and Three-Electron Systems* (Singapore: World Scientific)

[10] Oks E and Gavrilenko V P 1983 *Opt. Commun.* **46** 205

[11] Volod'ko D A, Gavrilenko V P and Oks E 1987 *18th Int. Conf. on Phenomena in Ionized Gases (Swansea, UK)* p 604

[12] Kadomtsev B B 1982 *Collective Phenomena in Plasma* (Oxford: Pergamon)

[13] Oks E 2019 *Atoms* **7** 25
[14] Oks E 2016 *J. Phys. B: At. Mol. Opt. Phys.* **49** 065701
[15] Zakatov L P, Plakhov A G, Shapkin V V and Sholin G V 1971 *Sov. Phys. Dokl.* **16** 451
[16] Karfidov D M and Lukina N A 1997 *Phys. Lett.* A **232** 443
[17] Gavrilenko V P 1996 *Pis'ma v Zh. Tech. Phys. (Sov. Phys. Tech. Phys. Lett.)* **22** 23 (in Russian)
[18] Oks E 2020 *Spectrochim. Acta* B **167** 105815
[19] Baranger M and Mozer B 1961 *Phys. Rev.* **123** 25
[20] Cooper W S and Ringler H 1969 *Phys. Rev.* **179** 226
[21] Dion D R and Hirshfelder J O 1976 *Adv. Chem. Phys.* **35** 265
[22] Gavrilenko V P and Oks E 1981 *Sov. Phys. JETP* **53** 1122 http://jetp.ras.ru/cgi-bin/e/index/e/53/6/p1122?a=list
[23] Gavrilenko V P and Oks E 1987 *Sov. J. Plasma Phys.* **13** 22
[24] Oks E, Böddeker St and Kunze H-J 1991 *Phys. Rev.* A **44** 8338
[25] Bertschinger G 1980 Messungen von VUV Linien an einem dichten Z-Pinch-Plasma *PhD Thesis* Ruhr University Bochum
[26] Gavrilenko V P, Faenov A Y, Magunov A I, Pikuz T A, Skobelev I Y, Kim K Y and Milchberg H M 2006 *Phys. Rev.* A **73** 013203
[27] Renner O, Dalimier E, Oks E, Krasniqi F, Dufour E, Schott R and Förster E 2006 *J. Quant. Spectrosc. Radiat. Transfer* **99** 439
[28] Woolsey N, Howe J, Chambers D, Courtois C, Förster E, Gregory C, Hall I, Renner O and Uschmann I 2007 *High Energy Density Phys.* **3** 292
[29] Oks E, Dalimier E and Faenov A Y *et al* 2017 *Opt. Express* **25** 1958
[30] Oks E, Dalimier E and Faenov A Y *et al* 2017 *J. Phys. B: At. Mol. Opt. Phys.* **50** 245006
[31] Zhuzhunashvili A I and Oks E 1977 *Sov. Phys. JETP* **46** 1122 http://jetp.ras.ru/cgi-bin/e/index/e/46/6/p1122?a=list
[32] Dalimier E, Oks E and Renner O 2014 *Atoms* **2** 178
[33] Dalimier E, Pikuz T A and Angelo P 2018 *Atoms* **6** 45
[34] Oks E and Rantsev-Kartinov V A 1980 *Sov. Phys. JETP* **52** 50 http://jetp.ras.ru/cgi-bin/e/index/e/52/1/p50?a=list
[35] Jian L, Shali X, Quingguo Y, Lifeng L and Yufen W 2013 *J. Quant. Spectrosc. Radiat. Transfer* **116** 41
[36] Oks E, Dalimier E and Faenov A Y *et al* 2014 *J. Phys. B: At. Mol. Opt. Phys.* **47** 221001
[37] Oks E, Dalimier E and Faenov A Y *et al* 2017 *J. Phys.: Conf. Ser.* **810** 012004
[38] Dalimier E, Oks E and Renner O 2017 *AIP Conf. Proc.* **1811** 190003
[39] Oks E 1978 *Sov. Astron. Lett.* **4** 223
[40] Dalimier E and Oks E 2018 *Atoms* **6** 60
[41] Oks E, Dalimier E and Angelo P 2019 *Spectrochim. Acta* B **157** 1
[42] Baranger M 1958 *Phys. Rev.* **111** 481
[43] Kolb A and Griem H R 1958 *Phys. Rev.* **111** 514
[44] Smith E, Cooper J and Vidal C 1969 *Phys. Rev.* **185** 140
[45] Vidal C, Cooper J and Smith E 1970 *J. Quant. Spectrosc. Radiat. Transf.* **10** 1011 1971**11** 263
[46] Smith E and Hooper C 1967 *Phys. Rev.* **157** 126
[47] Smith E 1968 *Phys. Rev.* **166** 102

[48] Lisitsa V S and Sholin G V 1972 *Sov. Phys. JETP* **34** 484 http://jetp.ras.ru/cgi-bin/e/index/e/34/3/p484?a=list
[49] Derevianko A and Oks E 1996 *Physics of Strongly Coupled Plasmas* ed W D Kraeft and M Schlanges (Singapore: World Scientific) p 286 p
[50] Oks E 2015 *J. Quant. Spectrosc. Radiat. Transfer* **152** 74
[51] Griem H R 1974 *Spectral Line Broadening by Plasmas* (Cambridge, MA: Academic) https://shop.elsevier.com/books/spectral-line-broadening-by-plasmas/griem/978-0-12-302850-1
[52] Cross R J Jr 1967 *J. Chem. Phys.* **46** 609
[53] Fox K 1968 *J. Phys. A (Proc. Phys. Soc.)* **1, Ser. 2** 124
[54] Chandrasekaran T and Wilkerson T W 1969 *Phys. Rev.* **181** 329
[55] Lebedev V S and Beigman I L 1998 *Physics of Highly Excited Atoms and Ions* (Berlin: Springer)
[56] Griem H R and Shen K Y 1961 *Phys. Rev.* **122** 1490
[57] Griem H R 1964 *Plasma Spectroscopy* (New York: McGraw-Hill)
[58] Oks E 2006 *Stark Broadening of Hydrogen and Hydrogenlike Spectral Lines in Plasmas: The Physical Insight* (Oxford: Alpha Science International)
[59] Sanders P and Oks E 2018 *J. Phys. Commun.* **2** 035033
[60] Baryshenkov F F and Lisitsa V S 1981 *Sov. Phys. JETP* **53** 471 http://jetp.ras.ru/cgi-bin/e/index/e/53/3/p471?a=list
[61] Bureyeva L A and Lisitsa V S 2000 *A Perturbed Atom: Astrophysics and Space Physics Reviews* (Boca Raton, FL: CRC Press)
[62] Ispolatov Y and Oks E 1994 *J. Quant. Specrosc. Radiat. Transfer* **51** 129
[63] Oks E, Derevianko A and Ispolatov Y 1995 *J. Quant. Specrosc. Radiat. Transfer* **54** 307
[64] Oks E 2006 *AIP Conf. Proc.* **874** 19
[65] Iglesias C A, Lebowitz J and McGowan D 1983 *Phys. Rev.* A **32** 1667
[66] Lisitsa V S 1977 *Sov. Phys. Usp.* **122** 603
[67] Sholin G V, Demura A V and Lisitsa V S 1973 *Sov. Phys. JETP* **37** 1057 http://jetp.ras.ru/cgi-bin/e/index/e/37/6/p1057?a=list
[68] Derevianko A and Oks E 1994 *Phys. Rev. Lett.* **73** 2079
[69] Oks E 2010 *Int. Rev. Atom. Mol. Phys.* **1** 169
[70] Oks E 2012 Atomic processes in basic and applied physics *Springer Series on Atomic, Optical and Plasma Physics* ed V Shevelko and H Tawara vol 68 (New York: Springer) 393 p
[71] Oks E 2017 *J. Phys.: Conf. Ser.* **810** 012006
[72] Oks E 2000 *J. Quant. Spectr. Radiat. Transfer* **65** 405
[73] Oks E 2002 *J. Phys. B: At. Mol. Opt. Phys.* **35** 2251
[74] Flih S A, Oks E and Vitel Y 2003 *J. Phys. B: At. Mol. Opt. Phys.* **36** 283
[75] Stambulchik E and Demura A V 2016 *J. Phys. B: At. Mol. Opt. Phys.* **49** 035701
[76] Sahal-Brechot S 1969 *Astron. Astrophys.* **1** 91
[77] Strekalov M L and Burshtein A I 1972 *Sov. Phys. JETP* **34** 53 http://jetp.ras.ru/cgi-bin/e/index/e/34/1/p53?a=list

Chapter 2

Nonlinear effects in the radiation of satellites of hydrogenic spectral lines from plasmas and their applications

2.1 Satellites under a one-dimensional one-mode monochromatic electric field in the non-opaque case

In this section, we consider a hydrogenic atom/ion that has a nuclear charge of Z under a linearly-polarized field $\mathbf{E}_0 \cos \omega t$, such as that of a laser field. As early as 1933, Blochinzew [1] analytically solved this problem for a model hydrogenic line consisting of just one Stark component. His result for the profile of the Stark component can be represented in the following way:

$$S_{\text{profile}}(\Delta\omega/\omega) = \sum_{p=-\infty}^{\infty} [J_p(X\varepsilon)]^2 \delta(\Delta\omega/\omega - p). \tag{2.1}$$

In equation (2.1), $\Delta\omega$ is the detuning from the unperturbed frequency ω_0 of the spectral line and $J_p(X\varepsilon)$ are the Bessel functions, where

$$\varepsilon = 3\hbar E_0/(2Zm_e e\omega), \quad X = (nq)_{\text{upper}} - (nq)_{\text{lower}}, \quad q = (n_1 - n_2). \tag{2.2}$$

In equation (2.2), m_e and e are the electron mass and charge, respectively; n is the principal quantum number; n_1 and n_2 are the parabolic quantum numbers; and the subscripts 'upper' and 'lower' relate to the upper and lower energy levels involved in the radiative transition.

The profile $S(\Delta\omega/\omega)$ is normalized to unity:

$$\int_{-\infty}^{\infty} S_{\text{profile}}(\Delta\omega/\omega) \mathrm{d}(\Delta\omega/\omega) = 1. \tag{2.3}$$

It can be seen that, generally speaking, the profile of the Stark component splits into an infinite number of satellites separated from the unperturbed frequency ω_0 of the spectral line by $p\omega$, where p is any positive or negative integer (the value $p = 0$ corresponds to the 'main line,' that is, to $\Delta\omega = 0$).

We focus on the nonlinear dependence of the satellite intensities on the energy density $E_0^2/(8\pi)$ of the laser field. For this purpose, it is convenient to let

$$a = (X\varepsilon)^2, \tag{2.4}$$

which is the scaled energy density of the laser field (see equation (2.2)). Equation (2.1) can now be rewritten as

$$S_{\text{profile}}(\Delta\omega/\omega) = \sum_{p=-\infty}^{\infty} \left[J_p(a^{1/2})\right]^2 \delta(\Delta\omega/\omega - p). \tag{2.5}$$

Thus, the dependence of the satellite intensity $S(a, p)$ on the scaled energy density of the laser field and on the satellite number is as follows:

$$S(a, p) = \left[J_p(a^{1/2})\right]^2. \tag{2.6}$$

Let us consider, for example, the intensity of the first satellite $S(a,1) = [J_1(a^{1/2})]^2$. At relatively small energy densities of the laser field ($a \ll 1$), the intensity linearly depends on the energy density:

$$S(a, 1) \approx a/4. \tag{2.7}$$

However, at $a \geqslant 1$, the dependence of the intensity on the scaled energy density of the laser field becomes nonlinear. This is illustrated by figure 2.1, in which the dashed

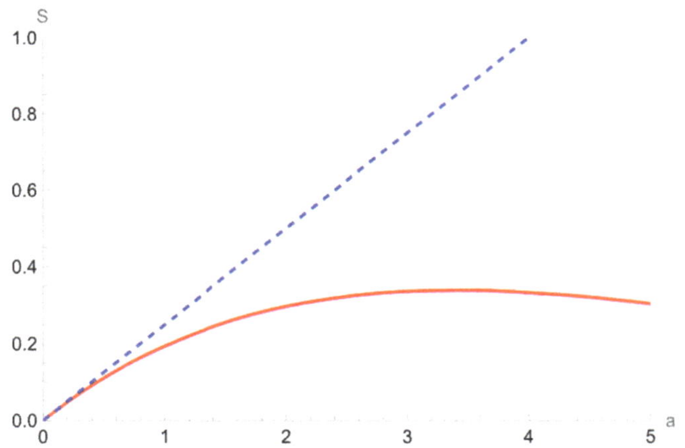

Figure 2.1. The intensity of the first satellite of the Blochinzew profile of the Stark component of a hydrogenic spectral line versus the scaled energy density of the laser field a, defined by equations (2.2) and (2.6). The solid line represents the actual dependence, while the dashed line represents a projected linear dependence, which is relevant only for $a \ll 1$.

line corresponds to an assumed linear dependence, while the solid line corresponds to the actual dependence.

Figures 2.2 and 2.3 present three-dimensional plots of the dependence of the satellite intensity on the scaled energy density a of the laser field and on the satellite number p. It can be seen that for any satellite, the dependence on the energy density of the laser field is highly nonlinear.

The same problem has sometimes been treated in the quasistatic limit, that is $\omega \to 0$. In this case, the profile mimics the distribution of the instantaneous field intensity:

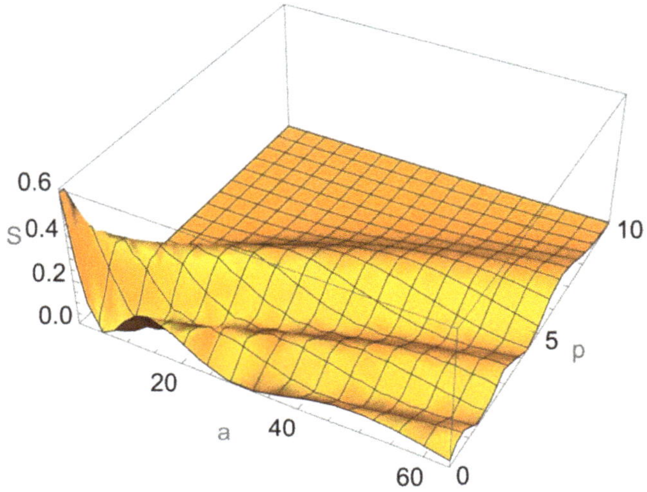

Figure 2.2. Dependence of the satellite intensity on the scaled energy density a of the laser field and on the satellite number p.

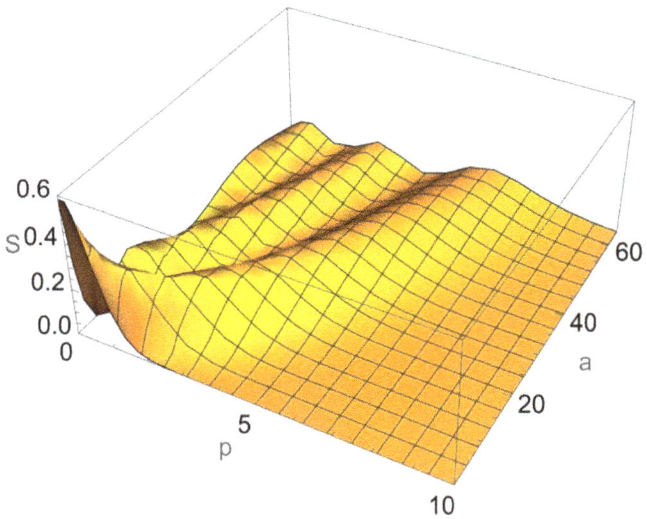

Figure 2.3. The same as in figure 2.2 but from the alternative viewpoint, so that together with figure 2.2, it gives a comprehensive view of the nonlinear dependence on the scaled energy density of the laser field.

$$S_{\text{quasist}}(p) = (1/\pi)[a - p^2]^{-1/2}. \tag{2.8}$$

Figures 2.4–2.8 provide comparisons between the quasistatic results from equation (2.8) with the actual satellite intensities from equation (2.6) for values of the scaled laser field energy density of 1, 4, 16, 64, and 256, respectively.

It can be seen that for all the above values of the scaled laser field energy density, the actual dependence of the satellite intensity on the satellite number exhibits much more structure than its quasistatic counterpart. In distinction to the latter, the actual dependence is oscillatory and thus 'more nonlinear' than in the quasistatic case.

Figure 2.9 shows the dependence of the intensity of the main line $S(a,0)$ on the scaled energy density of the laser field. It can be seen that this dependence is also highly nonlinear.

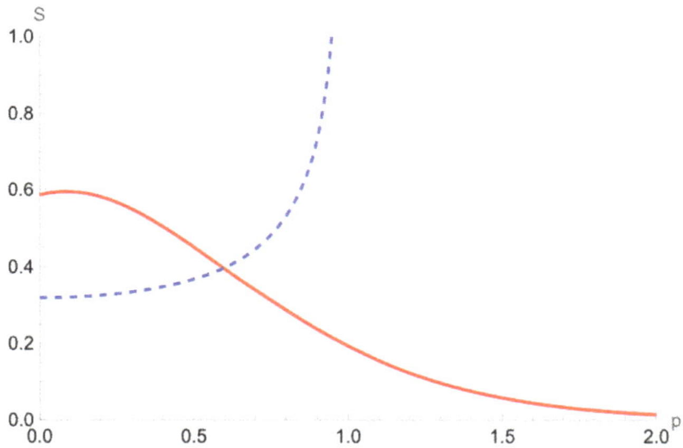

Figure 2.4. A comparison between the quasistatic result from equation (2.8), shown by the dashed line, and the actual satellite intensity from equation (2.6), shown by the solid line, for a scaled laser field energy density of $a = 1$.

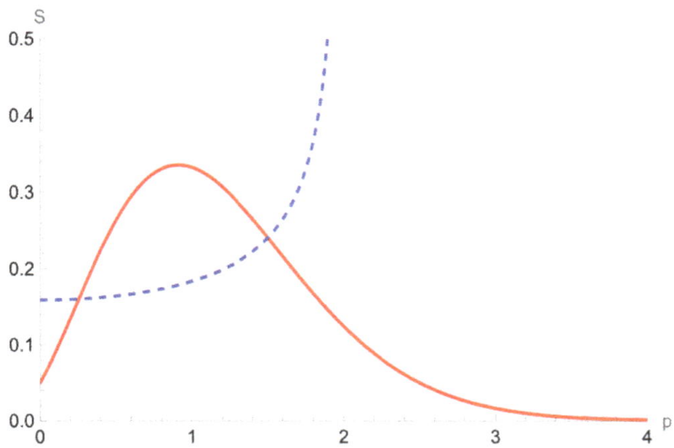

Figure 2.5. The same as in figure 2.4, but for $a = 4$.

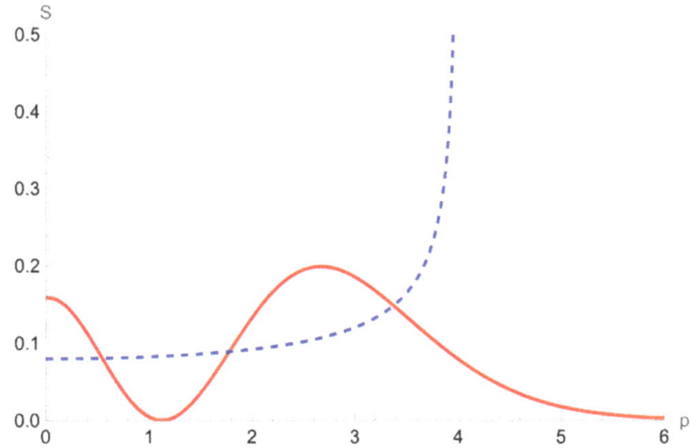

Figure 2.6. The same as in figure 2.4, but for a = 16.

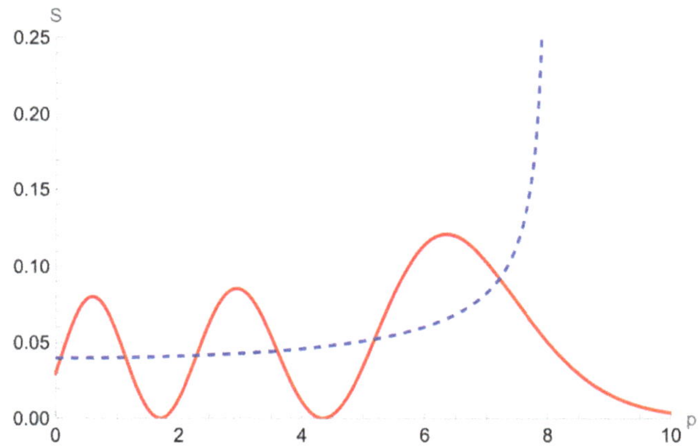

Figure 2.7. The same as in figure 2.4, but for a = 64.

2.2 Satellites under a one-dimensional one-mode monochromatic electric field in the opaque case

The satellites of hydrogenic spectral lines in laser-produced plasmas have not yet been used to measure the laser field (or the transverse electromagnetic field caused by the laser field)—to the best of our knowledge. Most probably, this is because the laser field would be deduced from the experimental ratio of the intensity of the satellite to the intensity of the main line. The problem would be that in plasmas produced by lasers, especially by relativistic laser pulses (those that drive plasma electrons to relativistic energies), the most frequently observed hydrogenic lines—Ly-alpha and Ly-beta—could have an opacity. In such a case, the observed peak intensity of the main line would be affected by the optical thickness, but the observed

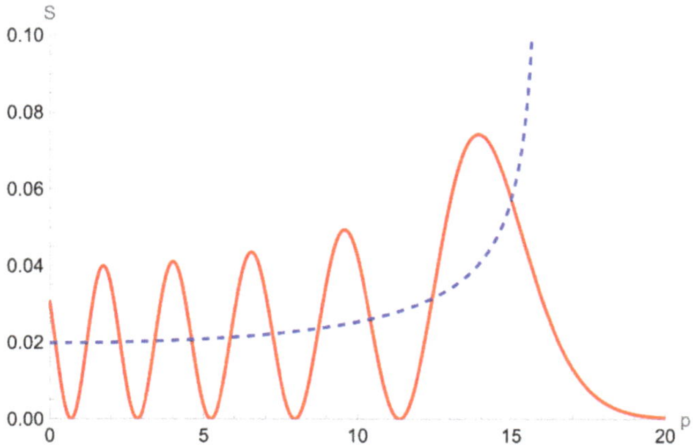

Figure 2.8. The same as in figure 2.4, but for a = 256.

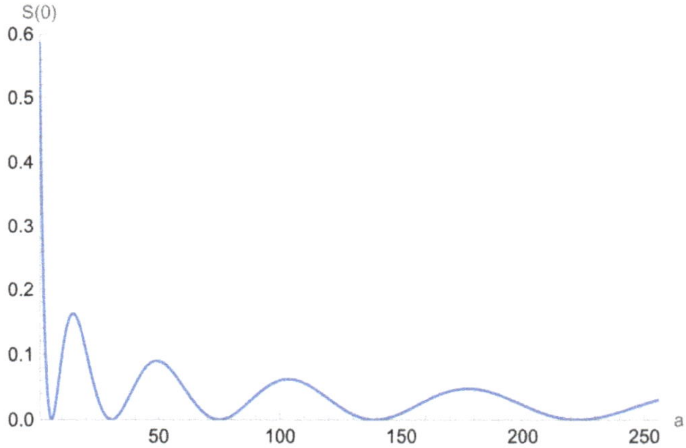

Figure 2.9. The dependence of the intensity of the main line $S(a,0)$ on the scaled energy density of the laser field.

peak intensity of the satellite would remain unaffected. Obviously, in this case, the theory presented in the previous section would not be a good tool for determining the laser field.

A method appropriate for such a situation was proposed in [2]. This method allows one to use hydrogenic lines that display satellites to measure both the laser field and the opacity.

In this section, we define the scaled dimensionless energy density of the laser field as

$$a = \varepsilon^2. \tag{2.9}$$

The profile of a multicomponent hydrogenic spectral line is as follows (see section 3.1 of [3]):

$$S(\Delta\omega/\omega) = \sum_{p=-\infty}^{+\infty} I(p, a^{1/2})\delta(\Delta\omega/\omega) - p,$$
$$I(p, \varepsilon) = \left[f_0\delta_{p0} + 2\sum_{k=1}^{k_{\max}} f_k J_p^2\left(X_k a^{1/2}\right)\right] / \left(f_0 + 2\sum f_k\right). \tag{2.10}$$

In equation (2.9), f_0 is the total intensity of all central Stark components and f_k is the intensity of the lateral Stark component with the number $k = 1, 2, \ldots, k_{\max}$.

The focus of [2] was the case in which $\varepsilon < 1$ and thus $a < 1$. In this case, the intensities of the satellites are significantly smaller than the intensity of the main line. Therefore, the satellites would be optically thin even if the optical depth τ_0 the main line were greater than unity.

The ratio of the intensity of the second satellite to the intensity of the first satellite has the form:

$$R_{21}\left(a^{1/2}\right) = I\left(2, a^{1/2}\right)/I\left(1, a^{1/2}\right). \tag{2.11}$$

This ratio depends only on the scaled energy density of the electric field. Therefore, from the experimental value of the ratio R_{21}, one can determine the value of a and then (using equations (2.2) and (2.9)) the laser amplitude E_0.

As for the ratio of the intensity of the first satellite to the intensity of the main line, at the zero optical depth of the main line, it would be

$$R_{10}\left(\tau_0 = 0, a^{1/2}\right) = I\left(1, a^{1/2}\right)/I\left(0, a^{1/2}\right). \tag{2.12}$$

Below, we illustrate the dependence of the ratio R_{21} on the scaled energy density of the laser field a and the corresponding dependence of the ratio R_{10}, calculated at $\tau_0 = 0$ for several hydrogenic lines in two different directions of observation: perpendicular to the laser field and parallel to the laser field.

Figure 2.10 shows the dependence of the ratio R_{21} on the scaled energy density of the laser field a (solid line) and the corresponding dependence of the ratio R_{10}, calculated at $\tau_0 = 0$ (dashed line), for the Ly-beta line for observation perpendicular to the laser field.

Figure 2.11 presents the dependence of the ratio R_{21} on the scaled energy density of the laser field a (solid line) and the corresponding dependence of the ratio R_{10}, calculated at $\tau_0 = 0$ (dashed line) for the Ly-beta line for observation parallel to the laser field.

Figure 2.12 shows the dependence of the ratio R_{21} on the scaled energy density of the laser field a (solid line) and the corresponding dependence of the ratio R_{10}, calculated at $\tau_0 = 0$ (dashed line) for the Ly-delta line for observation perpendicular to the laser field.

Figure 2.13 presents the dependence of the ratio R_{21} on the scaled energy density of the laser field a (solid line) and the corresponding dependence of the ratio R_{10}, calculated at $\tau_0 = 0$ (dashed line), for the Ly-delta line for observation parallel to the laser field.

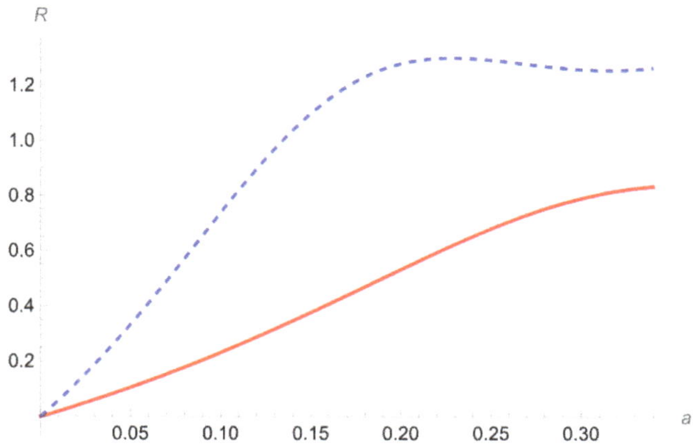

Figure 2.10. The dependence of the ratio R_{21} on the scaled energy density of the laser field a (solid line) and the corresponding dependence of the ratio R_{10}, calculated at $\tau_0 = 0$ (dashed line), for the Ly-beta line for observation perpendicular to the laser field.

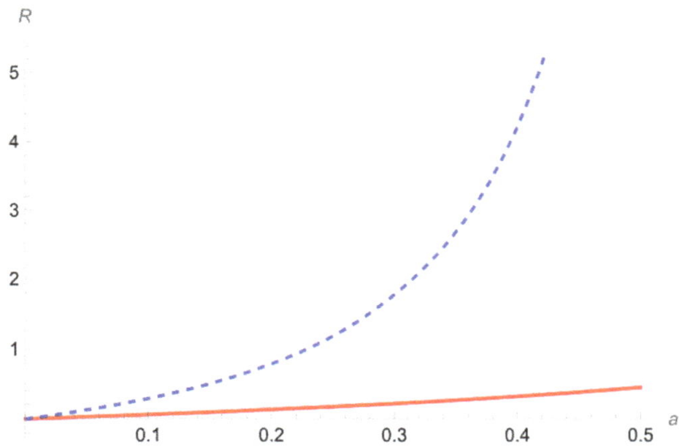

Figure 2.11. The dependence of the ratio R_{21} on the scaled energy density of the laser field a (solid line) and the corresponding dependence of the ratio R_{10}, calculated at $\tau_0 = 0$ (dashed line), for the Ly-beta line for observation parallel to the laser field.

Figure 2.14 shows the dependence of the ratio R_{21} on the scaled energy density of the laser field a (solid line) and the corresponding dependence of the ratio R_{10}, calculated at $\tau_0 = 0$ (dashed line), for the Balmer-beta line for observation perpendicular to the laser field.

Figure 2.15 presents the dependence of the ratio R_{21} on the scaled energy density of the laser field a (solid line) and the corresponding dependence of the ratio R_{10}, calculated at $\tau_0 = 0$ (dashed line), for the Balmer-beta line for observation parallel to the laser field.

Figure 2.16 shows the dependence of the ratio R_{21} on the scaled energy density of the laser field a (solid line) and the corresponding dependence of the ratio R_{10},

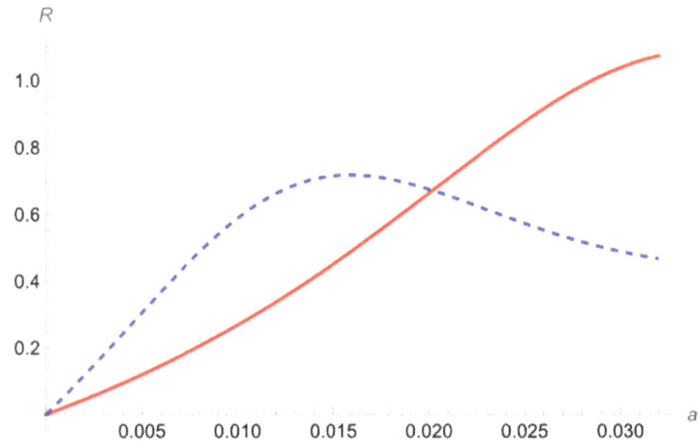

Figure 2.12. The dependence of the ratio R_{21} on the scaled energy density of the laser field a (solid line) and the corresponding dependence of the ratio R_{10}, calculated at $\tau_0 = 0$ (dashed line), for the Ly-delta line for observation perpendicular to the laser field.

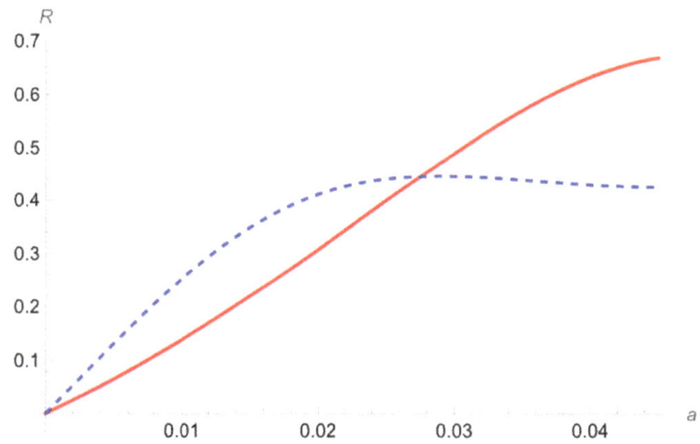

Figure 2.13. The dependence of the ratio R_{21} on the scaled energy density of the laser field a (solid line) and the corresponding dependence of the ratio R_{10}, calculated at $\tau_0 = 0$ (dashed line), for the Ly-delta line for observation parallel to the laser field.

calculated at $\tau_0 = 0$ (dashed line), for the Balmer-delta line for observation perpendicular to the laser field.

Figure 2.17 presents the dependence of the ratio R_{21} on the scaled energy density of the laser field a (solid line) and the corresponding dependence of the ratio R_{10}, calculated at $\tau_0 = 0$ (dashed line), for the Balmer-delta line for observation parallel to the laser field.

Figures 2.10–2.17 demonstrate the highly nonlinear dependence of these ratios on the scaled energy density of the laser field.

We now discuss the corresponding analytical results for cases of nonzero optical depth. At $\tau_0 > 0$ (and especially at $\tau_0 > 1$), the calculation of the ratio $R_{10}(\tau_0)$ has to

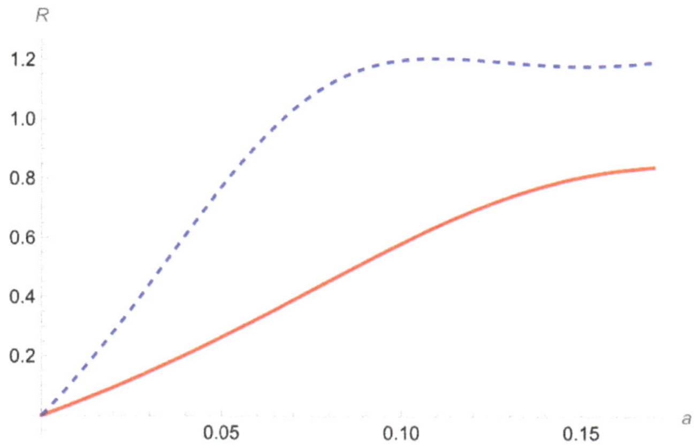

Figure 2.14. The dependence of the ratio R_{21} on the scaled energy density of the laser field a (solid line) and the corresponding dependence of the ratio R_{10}, calculated at $\tau_0 = 0$ (dashed line), for the Balmer-beta line for observation perpendicular to the laser field.

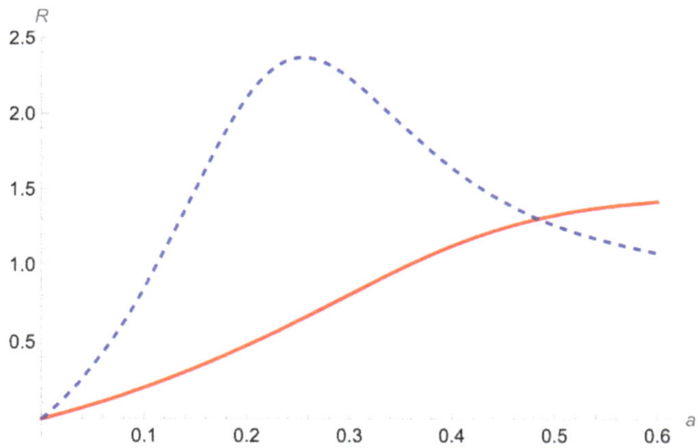

Figure 2.15. The dependence of the ratio R_{21} on the scaled energy density of the laser field a (solid line) and the corresponding dependence of the ratio R_{10}, calculated at $\tau_0= 0$ (dashed line), for the Balmer-beta line for observation parallel to the laser field.

be done in the following way [2]. The profile $P(\Delta\omega)$ of a spectral line, with an allowance for its opacity, is

$$P(\tau_0, \Delta\omega) = \{1 - \exp[-\tau_0 P_0(\Delta\omega)]\}/\tau_0. \tag{2.13}$$

In equation (2.13), $P_0(\Delta\omega)$ is the profile of the absorption coefficient with the normalization chosen to be $P_0(0) = 1$. At $\tau_0 = 0$, we have $P(0, \Delta\omega) = P_0(\Delta\omega)$; consequently, $P(0, 0) = 1$. The intensity at the peak of the normalized profile becomes

$$P(\tau_0, 0) = [1 - \exp(-\tau_0)]/\tau_0. \tag{2.14}$$

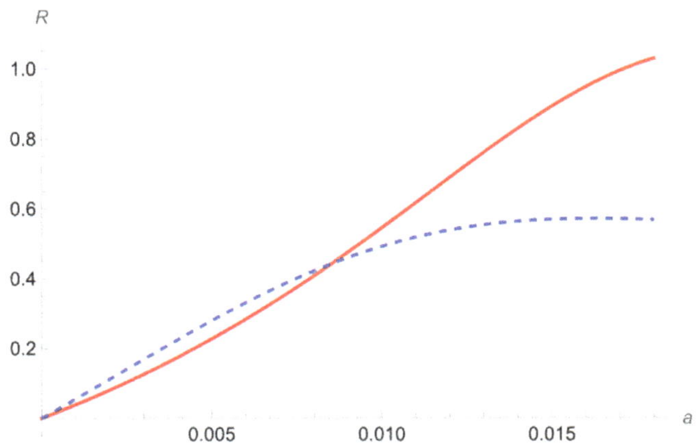

Figure 2.16. The dependence of the ratio R_{21} on the scaled energy density of the laser field a (solid line) and the corresponding dependence of the ratio R_{10}, calculated at $\tau_0 = 0$ (dashed line), for the Balmer-delta line for observation perpendicular to the laser field.

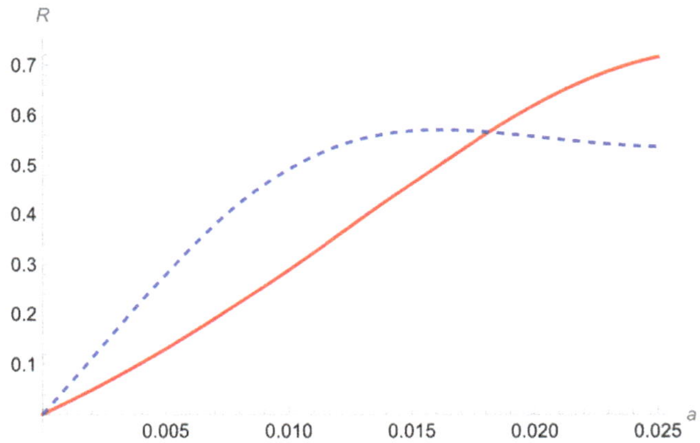

Figure 2.17. The dependence of the ratio R_{21} on the scaled energy density of the laser field a (solid line) and the corresponding dependence of the ratio R_{10}, calculated at $\tau_0 = 0$ (dashed line), for the Balmer-delta line for observation parallel to the laser field.

So, the factor reducing the intensity at the peak of the profile compared to the situation of the optically thin profile is

$$f(\tau_0) = [1 - \exp(-\tau_0)]/\tau_0. \tag{2.15}$$

Figure 2.18 shows a plot of the reducing factor f versus the optical depth τ_0. One can see that as the optical depth increases, the reduction of the peak intensity becomes more and more significant.

In view of the fact that the satellites are optically thin and that $P(0,0) = 1$, the ratio of the peak intensities of the second and first satellites R_{21} is still described by equation (2.11). However, for the ratio of the peak intensities of the first satellite and the main line, one finds that

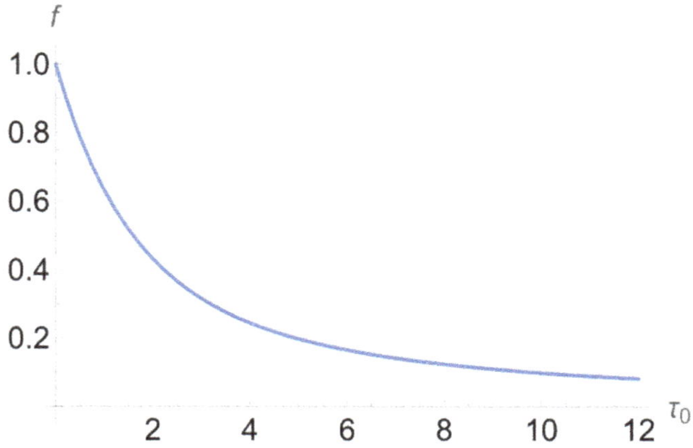

Figure 2.18. The reducing factor f from equation (2.15) versus the optical depth τ_0. Reproduced from [2]. CC BY 4.0.

$$R_{10}(\tau_0, a^{1/2}) = I(1, a^{1/2})/[I(0, a^{1/2})f(\tau_0)] \tag{2.16}$$

or

$$f(\tau_0) = I(1, a^{1/2})/[I(0, a^{1/2})R_{10}(\tau_0, a^{1/2})]. \tag{2.17}$$

So, the scaled energy density a (and thus the laser amplitude E_0) can be deduced from the experimental ratio R_{21}. Thereafter, on substituting the determined value of a and the experimental ratio R_{10} into the right-hand side of equation (2.16), it is possible to obtain the optical depth τ_0 of the main line. For the spectral lines Lyman-beta, Lyman-delta, Balmer-beta, and Balmer-delta, which are often utilized in plasma spectroscopy, the ratio $I(1,a)/I(0,a)$ that appears in equations (2.12) and (2.16) can be also deduced from figures 2.10–2.17.

2.3 Satellites under a one-dimensional two-mode monochromatic electric field

The corresponding profile of a Stark component, according to equation (3.2.8) of [3], is

$$S_{2,\ \mathrm{profile}}(\Delta\omega/\omega) = \sum_{p=-\infty}^{\infty} S(a, p)(\Delta\omega/\omega - p), \tag{2.18}$$

where a is defined by equation (2.24) and

$$S(a, p) = J_p^2(a^{1/2}) - J_{p-1}(a^{1/2})J_{p+1}(a^{1/2}). \tag{2.19}$$

For the first satellite ($p = 1$) at $a \ll 1$, the dependence of the satellite intensity on the scaled laser field energy density is approximately linear:

$$S(a, p) \approx a/8. \tag{2.20}$$

However, as the scaled energy density of the laser field grows, the corresponding dependence becomes nonlinear, as illustrated below.

Figure 2.19 shows the dependence of the intensity of the first satellite of the profile of the Stark component of a hydrogenic spectral line versus the scaled energy density of the two-mode laser field a.

Figures 2.20 and 2.21 present three-dimensional plots of the dependence of the satellite intensity on the scaled energy density a of the two-mode laser field and on

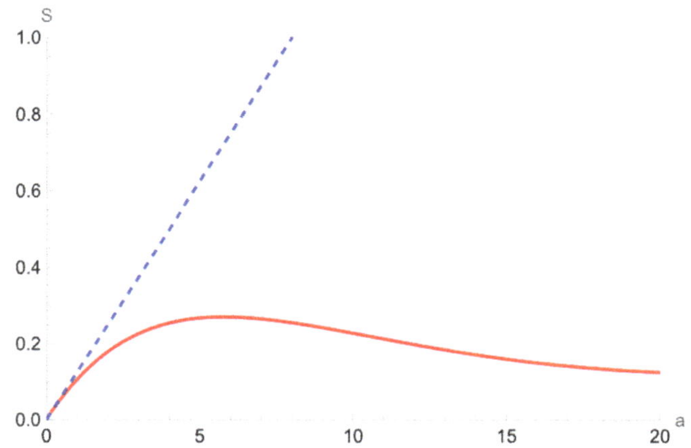

Figure 2.19. The dependence of the intensity of the first satellite of the profile of the Stark component of a hydrogenic spectral line versus the scaled energy density of the two-mode laser field a. The solid line represents the actual dependence, the dashed line represents the projected linear dependence, which is relevant only for $a \ll 1$.

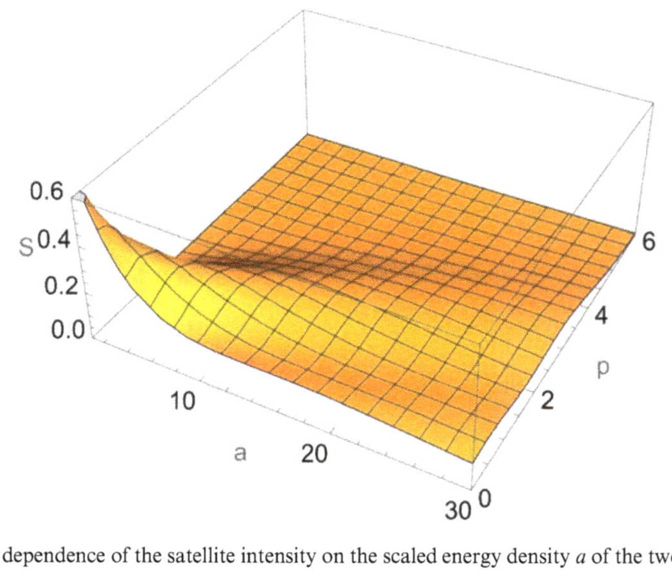

Figure 2.20. The dependence of the satellite intensity on the scaled energy density a of the two-mode laser field and on the satellite number p.

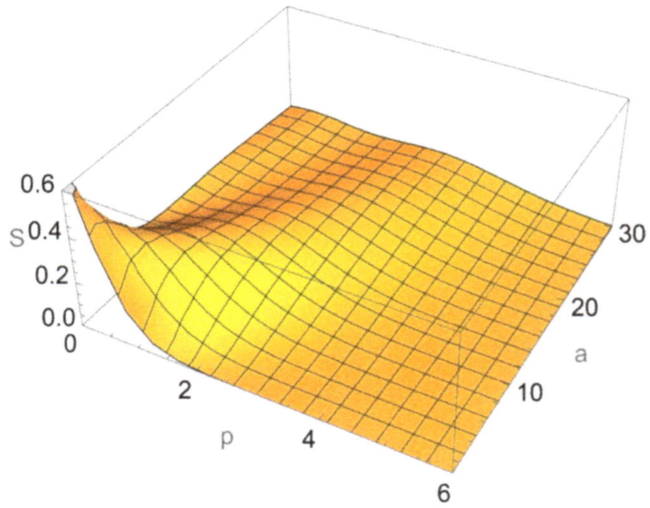

Figure 2.21. The same as in figure 2.20 but from the alternative viewpoint, so that together with figure 2.20, it gives the most comprehensive view of the nonlinear dependence on the scaled energy density of the two-mode laser field.

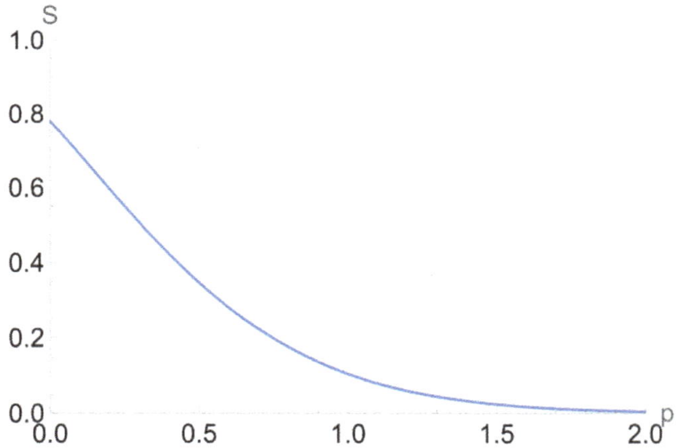

Figure 2.22. The dependence of the satellite intensity on the satellite number for the scaled energy density on the two-mode laser field a = 1.

the satellite number p. It can seen that for any satellite, the dependence on the energy density of the two-mode laser field is highly nonlinear.

Figure 2.22 shows the dependence of the satellite intensity on the satellite number for the scaled energy density on the two-mode laser field $a = 1$. It can be seen that the dependence is monotonic.

However, as the scaled energy density of the two-mode laser field increases, the dependence ceases to be monotonic. Figure 2.23 illustrates this for a = 4.

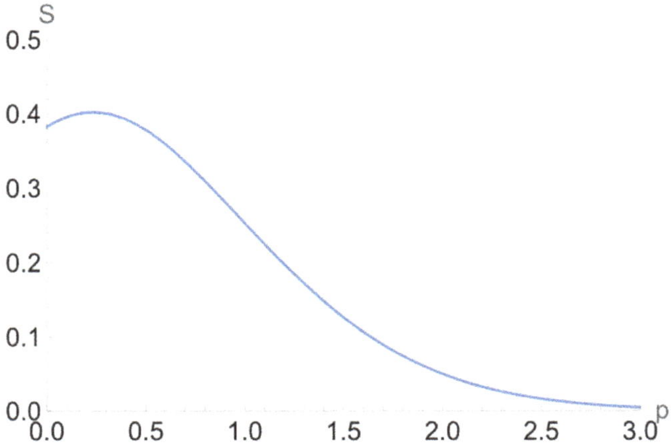

Figure 2.23. The same as in figure 2.22, but for a = 4.

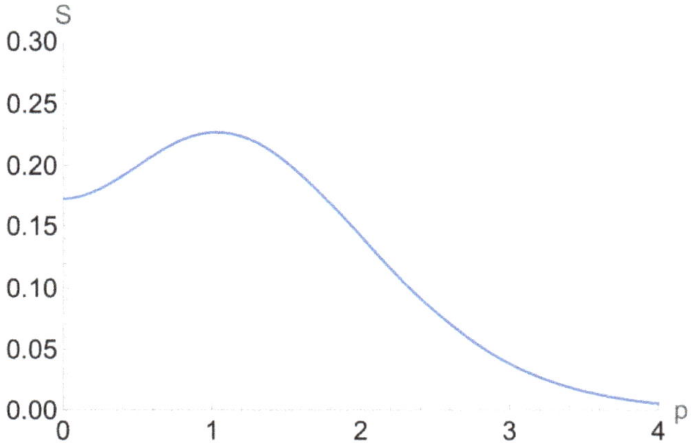

Figure 2.24. The same as in figure 2.22, but for a = 10.

Figure 2.24 presents the same kind of dependence for a = 10. It can be seen that for a = 10, the intensity of the first satellite exceeds the intensity of the main line.

Figures 2.25 and 2.26 show this dependence for a = 16 and a = 64, respectively. It can be seen that the intensities of the second and higher satellites never exceed the intensity of the main line—in contrast to the case of the one-mode laser field.

Figure 2.27 shows this kind of dependence for a = 256. It can be seen that as the scaled energy density of the two-mode laser field increases, the difference in the intensities of the various satellites decreases—in contrast to the case of the one-mode laser field.

Finally, figure 2.28 presents the dependence of the intensity of the main line on the scaled energy density of the two-mode laser field. It can be seen that the intensity oscillates much less than for the case of the one-mode laser field (see figure 2.9).

We note that all the dependencies presented in figures 2.22–2.28 are nonlinear.

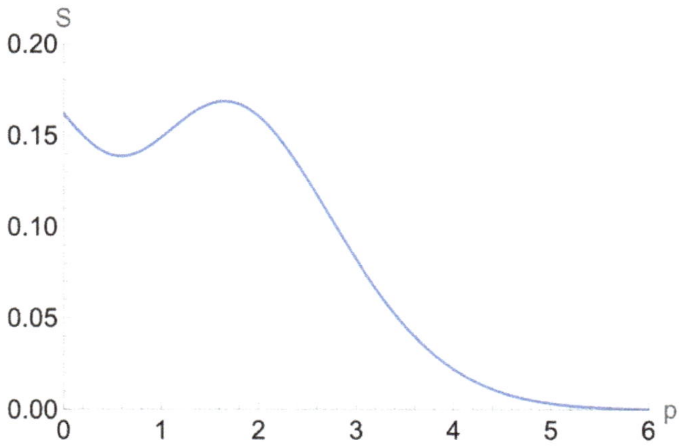

Figure 2.25. The same as in figure 2.22, but for $a = 16$.

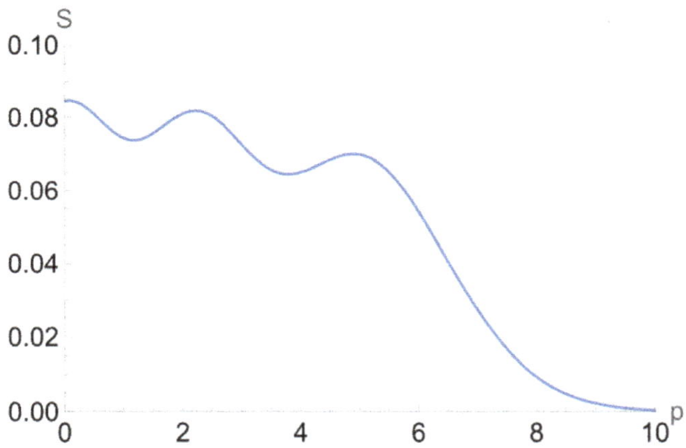

Figure 2.26. The same as in figure 2.22, but for $a = 64$.

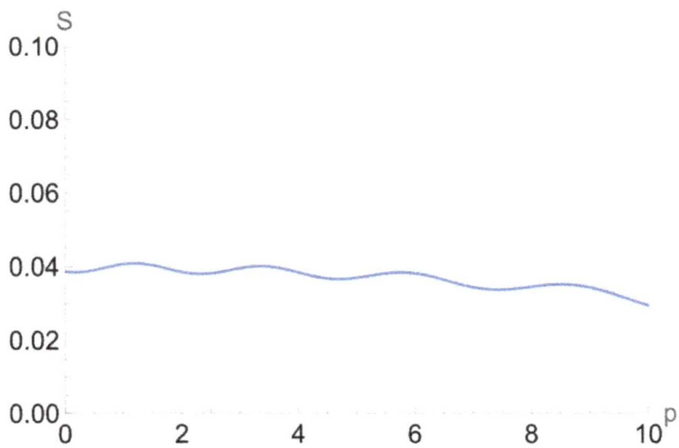

Figure 2.27. The same as in figure 2.22, but for $a = 256$.

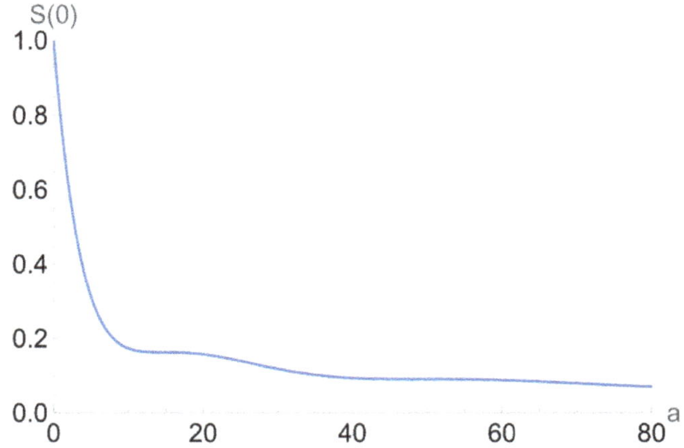

Figure 2.28. The dependence of the intensity of the main line on the scaled energy density of the two-mode laser field.

2.4 Satellites under a one-dimensional multimode monochromatic electric field

A one-dimensional multimode monochromatic electric field can be represented in the form

$$\mathbf{E}(t) = \sum_{j=1}^{N} \mathbf{E}_j \cos\left(\omega t + \varphi_j\right), \tag{2.21}$$

where the number of modes is $N \gg 1$. For this situation, the profile of a Stark component was calculated analytically by Lifshitz [4]:

$$S_{L,\,\text{profile}}(\Delta\omega/\omega) = \sum_{p=-\infty}^{\infty} S(a,\,p)(\Delta\omega/\omega - p), \tag{2.22}$$

where a is defined by equation (2.4) and

$$S(a,\,p) = I_{|p|}(a/2)\exp(-a/2). \tag{2.23}$$

In equation (2.23), $I_{|p|}(a/2)$ are the modified Bessel functions and the scaled energy density of the multimode monochromatic electric field is defined by equation (2.4) with the replacement of E_0 by

$$E_{0\text{L}} = \left(\sum_{j=1}^{N} E_j^2\right)^{1/2}. \tag{2.24}$$

For the first satellite ($p = 1$) at $a \ll 1$, the dependence of the satellite intensity of the scaled energy density of the field is approximately linear:

$$S(a,\,p) \approx a/4. \tag{2.25}$$

However, as the scaled energy density of the laser field grows, the corresponding dependence becomes nonlinear. This is illustrated by figure 2.29, where the dashed line corresponds to the projected linear dependence, while the solid line corresponds to the actual dependence.

Figures 2.30 and 2.31 present three-dimensional plots of the dependence of the satellite intensity on the scaled energy density a of the laser field and on the satellite number p. It can be seen that for any satellite, the dependence on the energy density of the laser field is highly nonlinear.

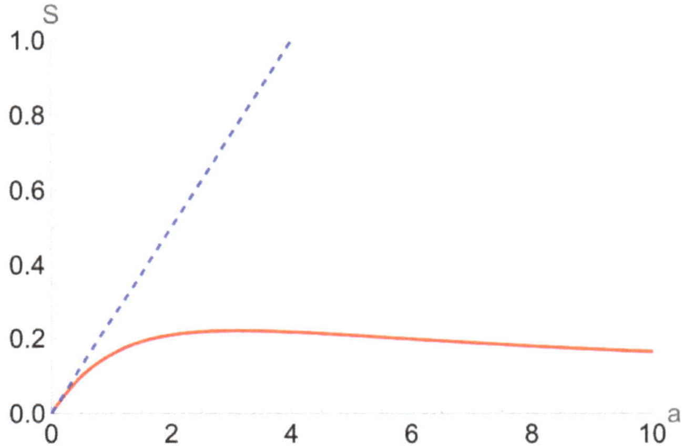

Figure 2.29. The intensity of the first satellite of the Lifshitz profile of the Stark component of a hydrogenic spectral line versus the scaled energy density of the multimode monochromatic electric field a. The solid line represents the actual dependence, the dashed line represents the projected linear dependence, which is relevant only for $a \ll 1$.

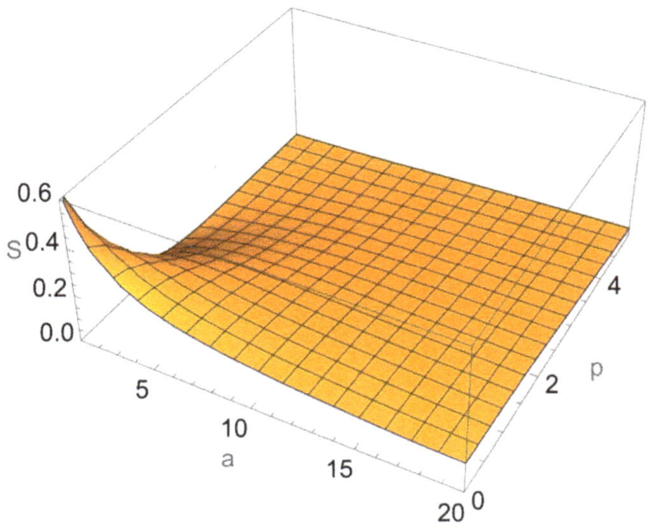

Figure 2.30. The dependence of the satellite intensity on the scaled energy density a of the multimode monochromatic electric field and on the satellite number p.

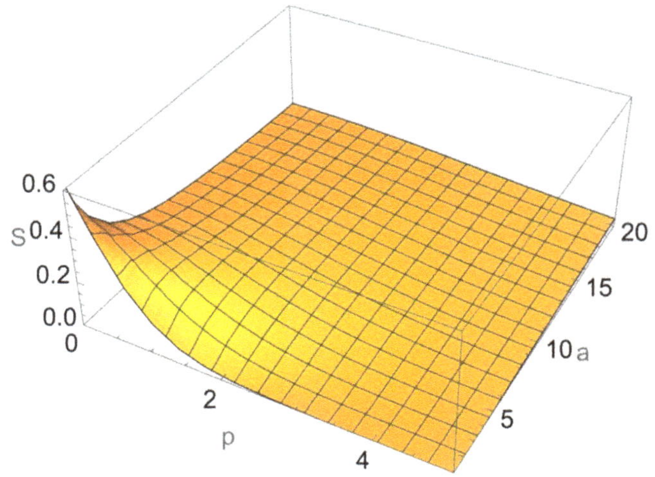

Figure 2.31. The same as in figure 2.30 but from the alternative viewpoint, so that together with figure 2.30, it gives the most comprehensive view of the nonlinear dependence on the scaled energy density of the multimode monochromatic electric field.

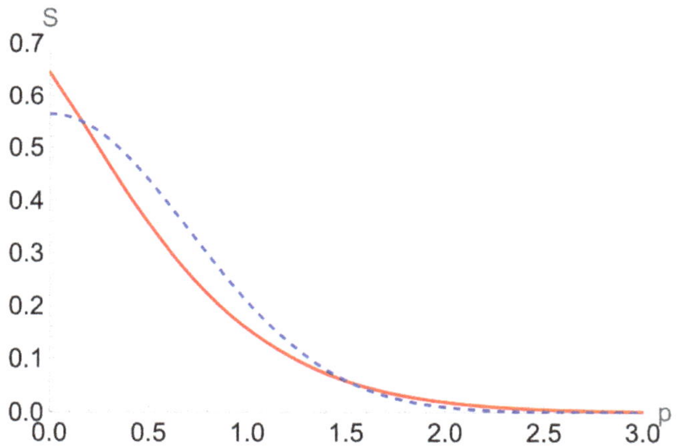

Figure 2.32. A comparison between the exact satellite intensity from equation (2.23), shown by the solid line, and the approximate result from equation (2.27), shown by the dashed line, for the scaled energy density of the multimode monochromatic electric field $a = 1$.

For the case in which

$$a \gg p \gg 1, \tag{2.26}$$

the satellite intensity from equation (2.23) simplifies to (see equation (3.2.4) from [3]):

$$S(a, p) \approx (\pi a)^{-1/2} \exp(-p^2/a). \tag{2.27}$$

It can be seen that at fixed a, the approximate envelope of the satellites is Gaussian.

Figures 2.32–2.35 present a comparison between the exact satellite intensity from equation (2.23), shown by the solid line, and the approximate result from equation

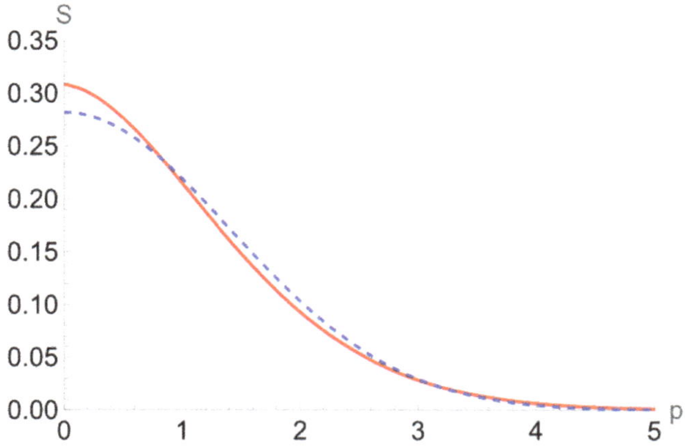

Figure 2.33. The same as in figure 2.32, but for $a = 4$.

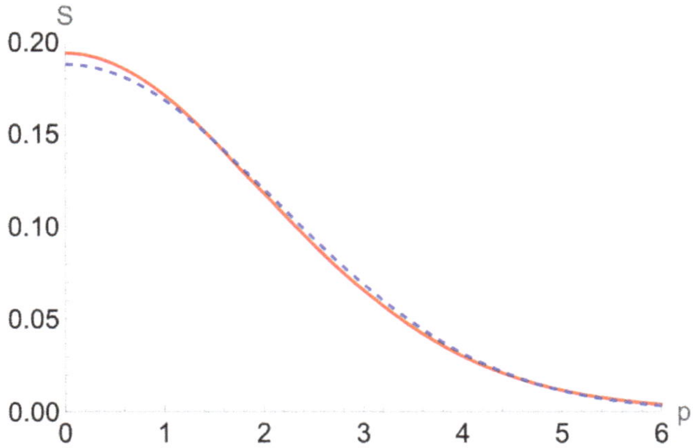

Figure 2.34. The same as in figure 2.32, but for $a = 9$.

(2.27), shown by the dashed line, for several values of the scaled energy density of the multimode monochromatic electric field.

From figures 2.32–2.35, one can see that at $a > 10$, the approximate satellite intensity from equation (2.27) practically coincides with the exact result from equation (2.23). It can also be seen that for the multimode electric field, the envelope of the satellites does not exhibit the oscillatory behavior—in contrast to the cases of the one-mode field and the two-mode field.

Finally, figure 2.36 presents the dependence of the intensity of the main line on the scaled energy density of the multimode monochromatic electric field. It can be seen that the intensity does not oscillate—in contrast to the cases of the one-mode field and the two-mode field (see figures 2.9 and 2.28).

We note that all the dependencies presented in figures 2.32–2.36 are nonlinear.

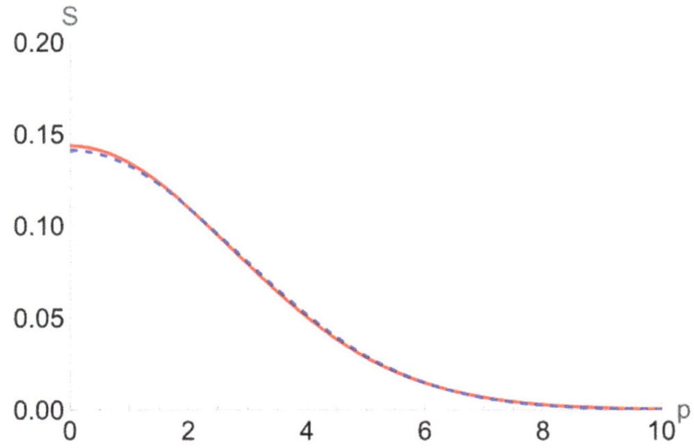

Figure 2.35. The same as in figure 2.32, but for $a = 16$.

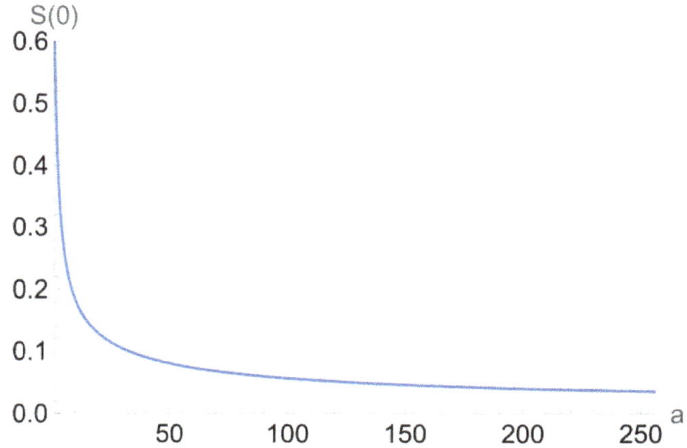

Figure 2.36. The dependence of the intensity of the main line on the scaled energy density of the multimode monochromatic electric field.

2.5 Satellites under a two-dimensional multimode electric field

In all the preceding sections of this chapter, the monochromatic electric field was considered to be one-dimensional. In this section, we consider the Stark broadening of hydrogenic spectral lines by a two-dimensional multimode monochromatic electric field. One of the primary differences between the effect of one-dimensional multimode monochromatic fields on hydrogenic spectral lines, described in the previous section, and the effect of the two-dimensional multimode monochromatic fields on hydrogenic spectral lines, presented in this section, involves the concept of *quasienergy states* introduced by Zeldovich [5] and Ritus [6], So, let us first briefly describe quasienergy states.

For an atom in a monochromatic electric field that has a frequency of ω, quasienergy states are those of the combined system 'atom plus field.' In the absence of a coupling between the atom and the field, each atomic state of energy E_i corresponds to an infinite number of states of the combined system of energies $E_{ij}^{(0)} = E_i + j\hbar\omega$, where j is any positive or negative integer or zero. In other words, the quasienergy state of $E_{ij}^{(0)}$ is the combination of the atomic state E_i and j quanta of the field. Here, the superscript (0) indicates that the coupling between the atom and the field has been disregarded.

When allowance is made for the coupling between the atom and the field, the energies E_{ij} of the combined system generally differ from the unperturbed values $E_{ij}^{(0)}$:

$$E_{ij} = E_{ij}^{(0)} + K_i. \tag{2.28}$$

The quantities K_i are called *quasienergies*.

In terms of the quasienergies, the outcome for a hydrogenic atom under the one-dimensional multimode monochromatic electric field, presented in the previous section, was zero quasienergies. In other words, the outcome was only the splitting of the hydrogenic line into satellites, but not a shift of the positions of all satellites and the main line—the shift that would be the same for all satellites and for the main line. In contrast, for a hydrogenic atom under a two-dimensional multimode monochromatic electric field, the quasienergies are generally nonzero, as presented below.

We consider the hydrogenic Ly-alpha line of a hydrogenic atom/ion that has a nuclear charge of Z acted upon by the following two-dimensional multimode monochromatic electric field:

$$\mathbf{E}(t) = E_x(t)\mathbf{e}_x + E_z(t)\mathbf{e}_z, \tag{2.29}$$

$$E_x(t) = \sum_k E_k \cos(\omega t + \varphi_k) = \rho_x \cos(\omega t + \alpha_x), \tag{2.30}$$

$$E_z(t) = \sum_p E_p \cos\left(\omega t + \varphi_p\right) = \rho_z \cos(\omega t + \alpha_z). \tag{2.31}$$

In equations (2.30) and (2.31), the phases α_x and α_z are uniformly distributed within the $(0,2\pi)$ interval. The quantities ρ_x and ρ_z, which are the amplitudes of the x- and z-components of the field, respectively, have the Rayleigh distributions

$$W(\rho_x) = \left(2\rho_x/\rho_{x0}^2\right)\exp\left(-\rho_x^2/\rho_{x0}^2\right), \quad W(\rho_z) = \left(2\rho_z/\rho_{x0}^2\right)\exp\left(-\rho_z^2/\rho_{z0}^2\right). \tag{2.32}$$

In equation (2.32), ρ_{x0} and ρ_{z0} are the root-mean-square values of the field amplitudes ρ_x and ρ_z, respectively. Below, we focus on the situation in which $\rho_x \ll \rho_z$.

At the fixed difference of the phases

$$\Delta = \alpha_z - \alpha_x, \tag{2.33}$$

the spectrum of the Ly-alpha line is as presented in section 4.1.5 of [3], as follows (here and below, the atomic units $\hbar = e = m_e$ are used):

$$I_x(\Delta\omega) = \delta(\Delta\omega - K) + \delta(\Delta\omega + K), \tag{2.34}$$

$$I_y(\Delta\omega) = 2\delta(\Delta\omega), \tag{2.35}$$

$$I_z(\Delta\omega) = \sum_{p=-\infty}^{\infty} \left\{ \begin{matrix} 2J_{2p+1}{}^2(v_z)\delta[\Delta\omega - (2p+1)\omega] \\ + J_{2p}{}^2(v_z)[\delta(\Delta\omega - 2p\omega - K) + \delta(\Delta\omega - 2p\omega + K)] \end{matrix} \right\}. \tag{2.36}$$

Here, the quantity K is the quasienergy:

$$K = -\omega v_x J_1(v_z) \sin(\Delta), \tag{2.37}$$

where $J_1(v)$ are the Bessel functions. In equations (2.36) and (2.37),

$$v_x = 3\rho_x/(Z\omega), \quad v_z = 3\rho_z/(Z\omega). \tag{2.38}$$

So, the two-dimensional multimode monochromatic electric field indeed leads to nonzero values $\pm K$ of the quasienergy—in contrast to the case of the one-dimensional multimode monochromatic field.

We introduce the scaled dimensionless energy densities of each component of the field as follows:

$$a_x = v_x{}^2, \quad a_z = v_z{}^2. \tag{2.39}$$

Figures 2.37 and 2.38 present three-dimensional plots of the dependence of the quasienergy K (in units of the field frequency ω) on the scaled energy densities a_x and a_z of the two components of the field for the phase difference $\Delta = \pi/2$.

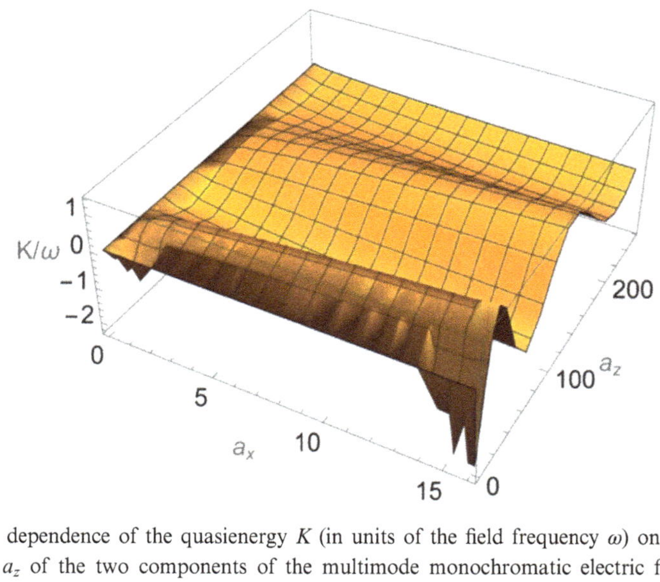

Figure 2.37. The dependence of the quasienergy K (in units of the field frequency ω) on the scaled energy densities a_x and a_z of the two components of the multimode monochromatic electric field for the phase difference $\Delta = \pi/2$.

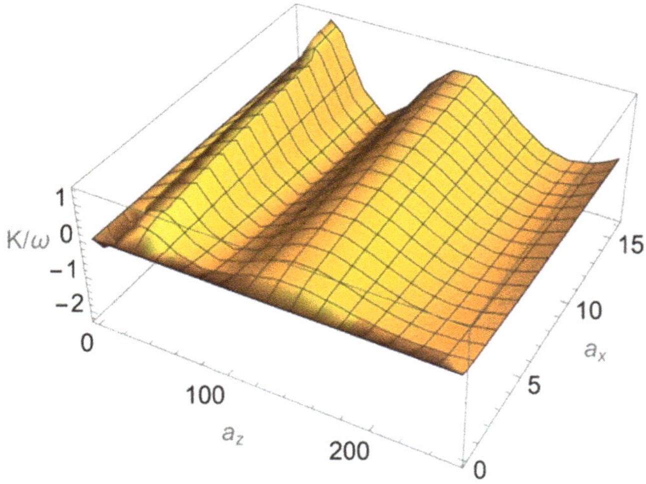

Figure 2.38. The same as in figure 2.37 but from the alternative viewpoint, so that together with figure 2.37, it gives the most comprehensive view of the nonlinear dependence of the quasienergy on the scaled energy densities a_x and a_z of the two components of the multimode monochromatic electric field for the phase difference $\Delta = \pi/2$.

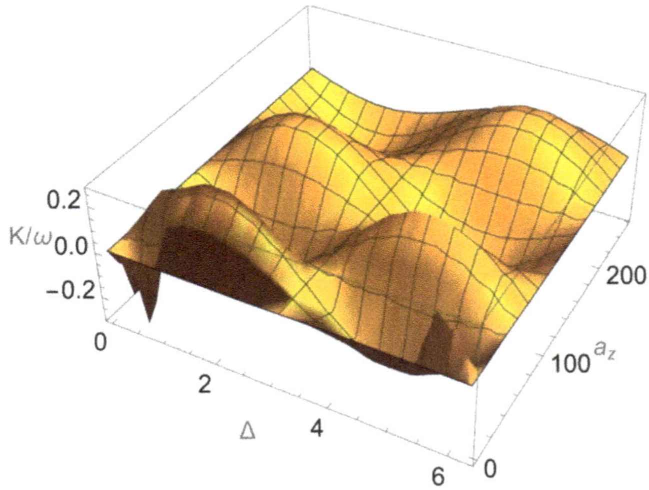

Figure 2.39. The dependence of the quasienergy K (in units of the field frequency ω) on the scaled energy density a_z of the z-component of the field and on the phase difference Δ for $a_x = 1$.

Figures 2.39 and 2.40 display three-dimensional plots of the dependence of the quasienergy K (in units of the field frequency ω) on the scaled energy density a_z of the z-component of the field and on the phase difference Δ for $a_x = 1$.

Figure 2.41 depicts the dependence of the quasienergy K (in units of ω) on the scaled energy density a_z of the z-component of the field for $a_x = 1$ and on the phase difference $\Delta = \pi/2$.

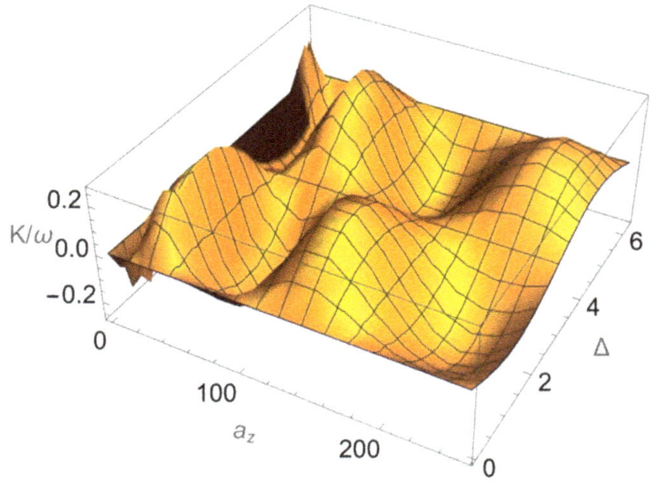

Figure 2.40. The same as in figure 2.39 but from the alternative viewpoint, so that together with figure 2.39, it gives the most comprehensive view of the nonlinear dependence of the quasienergy on the scaled energy density a_z of the z-component of the multimode monochromatic electric field and on the phase difference Δ for $a_x = 1$.

From figures 2.37–2.41, it can be seen that the dependence of the quasienergy on the energy density of the field components is highly nonlinear.

The next task is to average over the phase difference Δ (except that there is no need to average $I_y(\Delta\omega) = 2\delta(\Delta\omega)$). The result is as follows [7]:

$$\langle I_x(\Delta\omega)\rangle_{\text{phase}} = (2/\pi)\left\{\left[3\rho_x J_1(a_z^{1/2})/Z\right]^2 - (\Delta\omega)^2\right\}^{-1/2} \theta\left[3\rho_z \mid J_1(a_z^{1/2}) \mid /Z - |\Delta\omega|\right], \tag{2.40}$$

$$\begin{aligned}\langle I_z(\Delta\omega)\rangle = (1/\pi) \sum_{p=-\infty}^{\infty} \Big\{ & 2J_{2p+1}^2(a^{1/2})\big\{[3\rho_x J_1(a_z^{1/2})/Z]^2 \\ & - [\Delta\omega - (2p+1)\omega]^2\big\}^{-1/2} \theta[3\rho_x \mid J_1(a_z^{1/2}) \mid /Z - |\Delta\omega - (2p+1)\omega|] \\ & + 2J_{2p}^2(a_z^{1/2})[(3\rho_x J_1(a_z^{1/2})/Z)]^2 \\ & - (\Delta\omega - 2p\omega)^2\big\}^{-1/2} \theta[(3\rho_x \mid J_1(a_z^{1/2}) \mid /Z - \mid \Delta\omega - 2p\omega \mid)].\end{aligned} \tag{2.41}$$

In equations (2.40) and (2.41), $\theta(w)$ is the step function.

From equations (2.40) and (2.41), one can see that the Ly-alpha spectrum, averaged over the phase difference, comprises terms of the following kind (with different statistical weights):

$$S_0(D) = (1/\pi)\left\{\left[3\rho_x J_1(a_z^{1/2})/(Z\omega)\right]^2 - D^2\right\}^{-1/2} \theta\left[3\rho_x \mid J_1(a_z^{1/2}) \mid /(Z\omega) - |D|\right]. \tag{2.42}$$

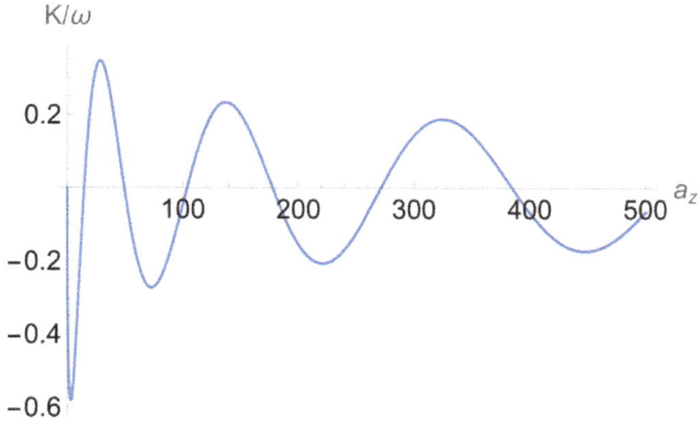

Figure 2.41. The dependence of the quasienergy K (in units of ω) on the scaled energy density a_z of the z-component of the field for the phase difference $\Delta = \pi/2$ and $a_x = 1$.

In equation (2.42),

$$D = (\Delta\omega)/\omega, \quad \text{or } D = (\Delta\omega)/\omega - (2p + 1), \quad \text{or } D = (\Delta\omega)/\omega - 2p. \tag{2.43}$$

Here, the selected unit of the frequency scale is the frequency ω of the two-dimensional multimode monochromatic electric field.

The next step is to integrate $S_0(D)$ from equation (2.42) over the distribution $W(\rho_x)$ from equation (2.32). The corresponding analytical result is:

$$\langle S_0(D)\rangle_{\rho x} = \exp\left\{-D^2/\left[3\rho_{x0}J_1(v_z)/(Z\omega)\right]^2\right\}/\left[\pi^{1/2}3\rho_{x0} \mid J_1(v_z) \mid/(Z\omega)\right]. \tag{2.44}$$

Finally, $\langle S_0(D)\rangle_{\rho x}$ from equation (2.44) has to be averaged over the distribution $W(\rho_z)$ from equation (2.32):

$$S(D) = \left\langle\langle S_0(D)\rangle_{\rho x}\right\rangle_{\rho z} = (2/\pi^{1/2})\int_0^{\infty} \mathrm{d}w w$$
$$\exp\{-w^2 - D^2/[aJ_1(bw)]^2\}/[a|J_1(bw)|]. \tag{2.45}$$

In equation (2.54),

$$w = \rho_z/\rho_{z0}, \quad a = 3\rho_{x0}/(Z\omega), \quad b = 3\rho_{z0}/(Z\omega). \tag{2.46}$$

In the case where

$$b < <1, \; ab/2 \ll \mid D \mid \ll a/(2b), \tag{2.47}$$

and assuming that the argument of the exponential in equation (2.45) has a maximum at some value $w = w_0$, where $bw_0 \ll 1$, the integration in equation (2.45) can be performed using the approximate analytical method of the steepest descent. The outcome is [7]:

$$S_0(D) \approx 2\exp[-4 \mid D \mid/(ab)]/(ab). \tag{2.48}$$

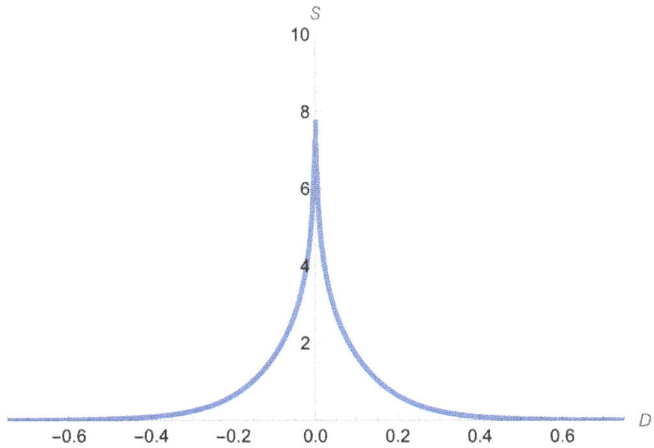

Figure 2.42. A Stark profile $S(D)$ calculated using equation (2.45) for $a = 0.75$, $b = 7.5$, where $D = (\Delta\omega)/\omega$, $a = 3\rho_{x0}/(Z\omega)$, $b = 3\rho_{z0}/(Z\omega)$.

Equation (2.48) shows that the Stark profile $S_0(D)$ of each satellite and of the main line can have a cusp in its center. Figure 2.42 presents a numerical example, calculated using equation (2.45), that confirms this analytical result.

The cusp-shaped type of spectral profile of the main line and of the satellites is a counterintuitive result. The prospective application of this result is the spectroscopic diagnosis of multidimensional Bernstein–Greene–Kruskal modes in laboratory and space plasmas [8].

It should be noted that plasmas in which the quasimonochromatic electric field dominates other broadening mechanisms are encountered, for instance, in flare stars (with an electron density $N_e \sim 10^{15}$ cm^{-3}), in solar flares ($N_e \sim 10^{13}$ cm^{-3}), as well as in the edge plasmas of tokamaks ($N_e \sim 10^{13}$–10^{15}) cm^{-3}—see, e.g. [9] and references therein.

2.6 Satellites in an elliptically polarized electric field

The splitting of hydrogenic spectral lines under an elliptically polarized laser field has been studied, for example, in [10, 11]. Here we present some results related to the situation in which the laser frequency is much greater than the corresponding atomic frequencies.

The elliptically polarized laser field can be represented in the form

$$\mathbf{E}(t) = \mathbf{e}_x b\varepsilon \sin(\omega t) + \mathbf{e}_z \varepsilon \cos(\omega t). \tag{2.49}$$

In equation (2.49), $\mathbf{e}_x$ and $\mathbf{e}_z$ are the unit vectors along the corresponding axes, b is the ellipticity degree, and

$$\varepsilon = E_0/(1 + b^2)^{1/2}, \tag{2.50}$$

where E_0 is the laser intensity. The instantaneous rotational frequency of the laser field is

$$\Omega(t) = b\omega/[\cos^2(\omega t) + b^2 \sin^2(\omega t)]. \tag{2.51}$$

The following results for the quasienergies were produced by employing the general formalism for the action of a high-frequency oscillatory field on a quantum system presented in section 2.3 of [3]. For the Ly-alpha line, the quasienergies are

$$w_{1,2} = 0, \quad w_3 = \kappa, \quad w_4 = -\kappa, \tag{2.52}$$

where

$$\kappa = b\omega v J_1(v), \quad v = 3E_0/[Z\omega(1 + b^2)^{1/2}]. \tag{2.53}$$

In equation (2.53), $J_1(v)$ is the Bessel function. (Here and below, atomic units are used.) It can be seen that the quasienergies λ_3 and λ_4 are nonlinear functions of the laser amplitude E_0, so they are nonlinear functions of the energy density (proportional to E_0^2) of the laser field.

For an arbitrary hydrogenlike spectral line, the analytical calculation of the quasienergies w in the elliptically polarized laser field can be reduced to calculating the Stark splitting in some effective static electric field, yielding

$$w = qb[a/(1 + b^2)]^{1/2} J_1\{[a/(1 + b^2)]^{1/2}\}. \tag{2.54}$$

In equation (2.54), the electric quantum number q is as defined in equation (2.2), and the scaled dimensionless energy density of the laser field, denoted by a, is:

$$a = [3nE_0/2\omega]^2. \tag{2.55}$$

Figures 2.43 and 2.44 show three-dimensional plots of the quasienergy w from equation (2.54) for $q = 1$ versus the scaled energy density a of the laser field and the ellipticity degree b.

Figure 2.45 presents the dependence of the quasienergy w from equation (2.54) for $q = 1$ on the scaled energy density a of the laser for three values of the ellipticity degree: $b = 0.2$ (dotted line), $b = 0.5$ (dashed line), and $b = 0.8$ (solid line).

From figures 2.43–2.45, it can be seen that the dependence of the quasienergies on the scaled energy density of the elliptically polarized laser field and on the ellipticity degree is highly nonlinear.

2.7 Nonlinear spectral effects in the case of Langmuir solitons

A set of Langmuir solitons separated by a distance L has the following spatial and temporal dependence [12]:

$$F(x, t) = E(x)\cos\omega t, \quad E(x) = E_0/ch(x/\lambda), \quad \lambda \ll L, \tag{2.56}$$

where

$$\omega = \omega_{pe} - 3T_e/\left(2m_e\omega_{pe}\lambda^2\right). \tag{2.57}$$

In equations (2.56) and (2.57), λ is the characteristic size of the soliton and ω_{pe} is the plasma electron frequency. The experimental task of identifying solitons consists of

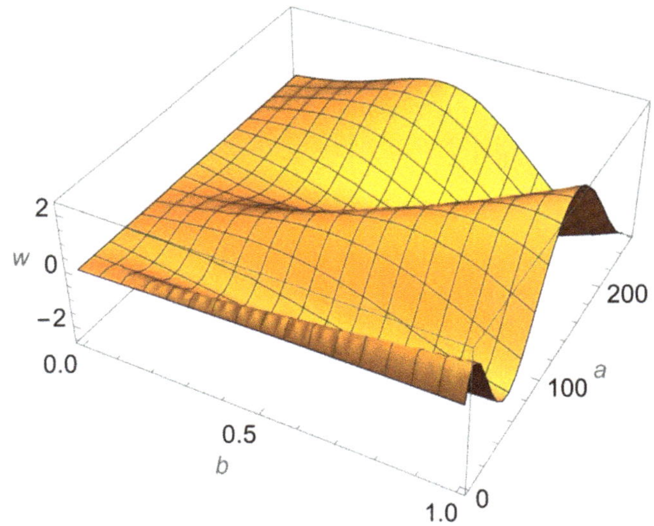

Figure 2.43. The quasienergy w from equation (2.54) for $q = 1$ versus the scaled energy density a of the laser field and the ellipticity degree b.

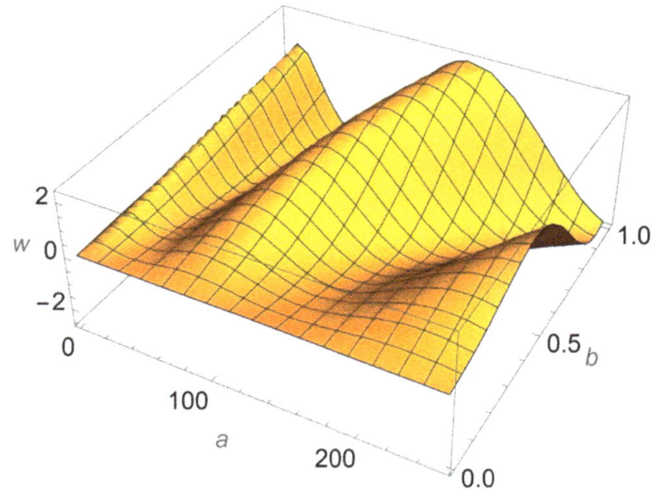

Figure 2.44. The same as in figure 2.43 but from the alternative viewpoint, so that together with figure 2.43, it gives the most comprehensive view of the nonlinear dependence of the quasienergy on the scaled energy density a of the laser field and on the ellipticity degree b.

detecting an electric field that oscillates at the frequency $\sim\omega_{pe}$ and confirming that the amplitude is spatially distributed according to the form factor $E(x)$ from equation (2.56). The corresponding method was developed in [13], which we follow here.

At a given coordinate x, the spectrum of a Stark component of a hydrogenic spectral line is described by equation (2.1) with the notations defined in equation (2.2). The next step is to average the spectrum over the form factor $E(x)$ from

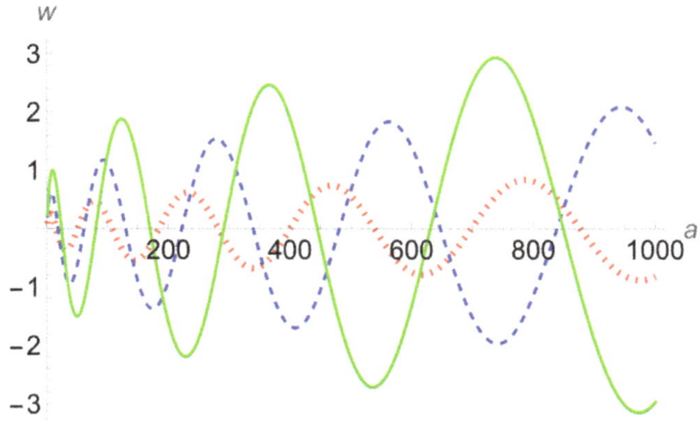

Figure 2.45. The dependence of the quasienergy w from equation (2.54) for $q = 1$ on the scaled energy density a of the laser for three values of the ellipticity degree: $b = 0.2$ (dotted line), $b = 0.5$ (dashed line), and $b = 0.8$ (solid line).

equation (2.56). First, we substitute the form factor in the argument of the Bessel function in equation (2.1). We then integrate with respect to x from $-\lambda$ to λ, or equivalently

$$f(p, a, b) = (1/b)\int_{-b/2}^{b/2} \mathrm{d}y J_p^2[a^{1/2}/ch(y)]. \quad (2.58)$$

In equation (2.58),

$$b = L/\lambda, \quad y = x/\lambda. \quad (2.59)$$

The scaled dimensionless energy density of the Langmuir wave, denoted by a, was given by equation (2.4) (see also equation (2.2)).

For $a \gg 1$, the maximum intensity has the satellite at the position $p_{\max}(a)$, where

$$p_{\max}(a) = a^{1/2} - 0.809(a)^{1/6}. \quad (2.60)$$

The spatially integrated intensity of the maximum intensity satellite is $I = f\left[p_{\max}(a), a, b\right]$.

Figure 2.46 demonstrates the dependence of the quantity I on the scaled separation b of the Langmuir solitons in the sequence for three values of the scaled energy density of the Langmuir wave: $a = 100$, $a = 225$, and $a = 400$. From figure 2.46, one can see that the integrated intensity of the most intense satellite decreases as either the parameter b increases (i.e. the separation L between Langmuir solitons in the sequence increases) or the scaled energy density of the Langmuir wave a increases.

Figure 2.47 displays a three-dimensional plot of the dependence of the spatially integrated intensity of the most prominent satellite on both the scaled energy density of the Langmuir wave a and on the scaled separation b of the Langmuir solitons in the sequence. The nonlinear character of the dependence of the intensity of this satellite on the scaled energy density of the Langmuir wave can clearly be seen.

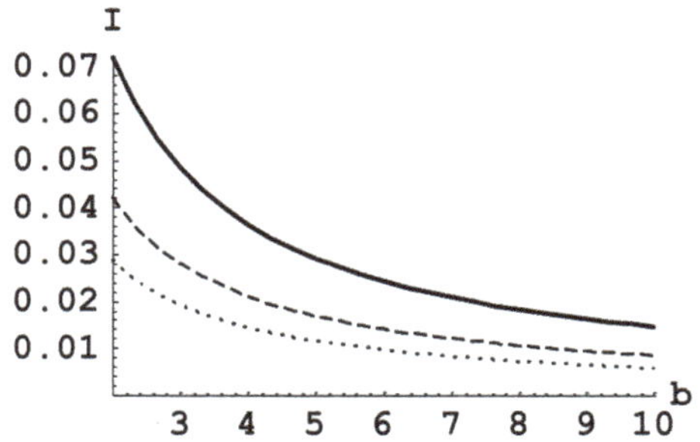

Figure 2.46. The spatially integrated intensity I of the most prominent satellite versus the scaled separation $b = L/\lambda$ of Langmuir solitons in the sequence for three values of the scaled amplitude $a = 3\hbar X_k E_0/(2Z_r m_e e\omega)$ of the soliton electric field: $a = 100$ (solid line), $a = 225$ (dashed line), and $a = 400$ (dotted line). Reproduced from [13]. CC BY 4.0.

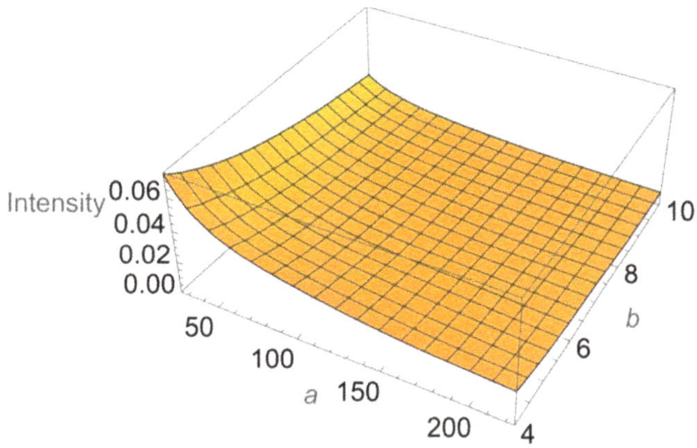

Figure 2.47. The spatially integrated intensity of the most prominent satellite versus the scaled energy density a of the soliton electric field and the scaled separation $b = L/\lambda$ of Langmuir solitons in the sequence. Reproduced from [13]. CC BY 4.0.

Figures 2.48 and 2.49 present the dependence of the ratio $f[p_{max}(a),a,b]/f[0,a,b]$ on a and b. The ratio presented in these figures is the spatially integrated intensity of the most intense satellite divided by the spatially integrated intensity of the main line (the 'zeroth' satellite). One can see that this ratio is also a nonlinear function of the scaled energy density a of the soliton electric field.

The next few figures present calculated profiles of the Ly-beta line for the direction of observation perpendicular to the soliton electric field $\mathbf{E}_0$. Figure 2.50 shows the profile of the Ly-beta line versus the scaled distance $\Delta\omega/\omega$ from the unperturbed position of this line for the (differently) scaled energy density of the Langmuir wave $\varepsilon = [3\hbar E_0/(2Z_r m_e e\omega)]^2 = 1$ and the scaled separation between Langmuir solitons in the sequence $b = L/\lambda = 2$ (solid line). The corresponding profile of the Ly-beta line for the situation in which the Langmuir waves are non-solitonic, for the same value of $\varepsilon = 1$, is shown by the dashed line.

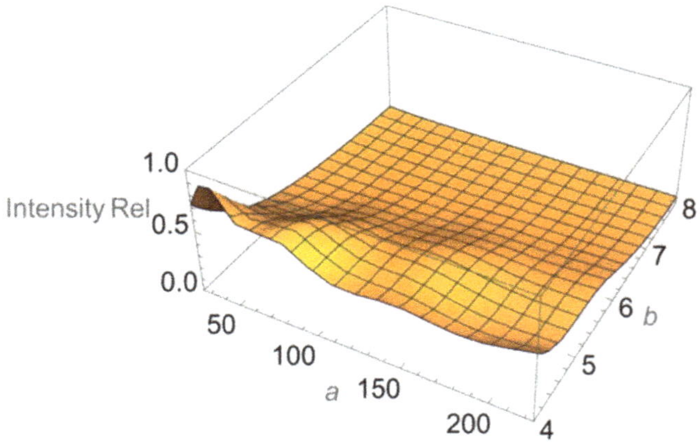

Figure 2.48. The ratio $f[p_{max}(a),a,b]/f[0,a,b]$ of the spatially integrated intensity of the most prominent satellite to the spatially integrated intensity of the main line (the 'zeroth' satellite) versus the scaled energy density of the soliton electric field and the scaled separation $b = L/\lambda$ of Langmuir solitons in the sequence. Reproduced from [13]. CC BY 4.0.

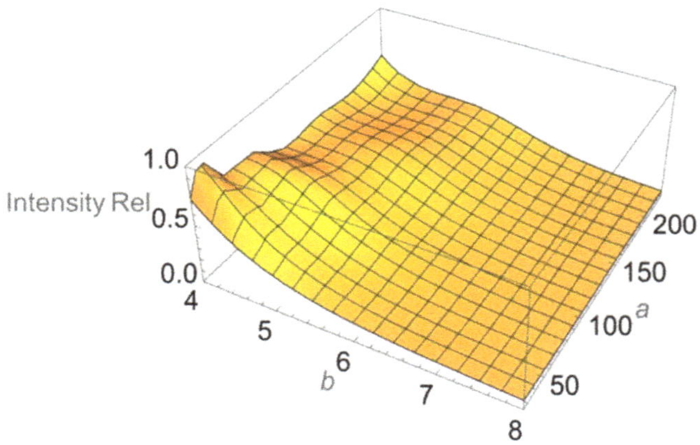

Figure 2.49. The same as in figure 2.48 but from the alternative viewpoint, so that together with figure 2.48, it gives the most comprehensive view of the nonlinear dependence of the intensity ratio under consideration on the scaled energy density of the soliton electric field a.

Paper [13] noted:

> 'The profiles are continuous (rather than being a set of satellites isolated from each other) because additional broadening mechanisms (the Stark broadening by plasma microfields and the Doppler broadening) were taken into account in amount of $\delta\omega = 2\omega$. An example of the fulfillment of the latter relation could be plasmas of multi-charged ions produced by a powerful Nd-glass laser, where at the surface of the critical density, the electron density is $N_e = 10^{21}$ cm^{-3} (or slightly higher due to relativistic effects) and the temperature would be up to $T \sim 10^3$ eV. It is seen that in

the case of the solitons, the profile is narrower than in the non-solitonic case. It is also seen that both profiles have practically the bell-shape without any significant features.'

Figure 2.51 depicts the same results as figure 2.50 but for more intense Langmuir solitons, corresponding to $\varepsilon = 9$. It can be seen that in the non-solitonic situation, shown by the dashed line, the profile has four maxima. In the solitonic case, the two secondary maxima become just shoulders.

Figure 2.52 presents the same results as figure 2.51 but for even more intense Langmuir solitons, corresponding to $\varepsilon = 36$. Both the solitonic profile (solid line) and the non-solitonic profile (dashed line) have six maxima. Some of these maxima are more pronounced in the solitonic situation, while other maxima are more pronounced in the non-solitonic situation.

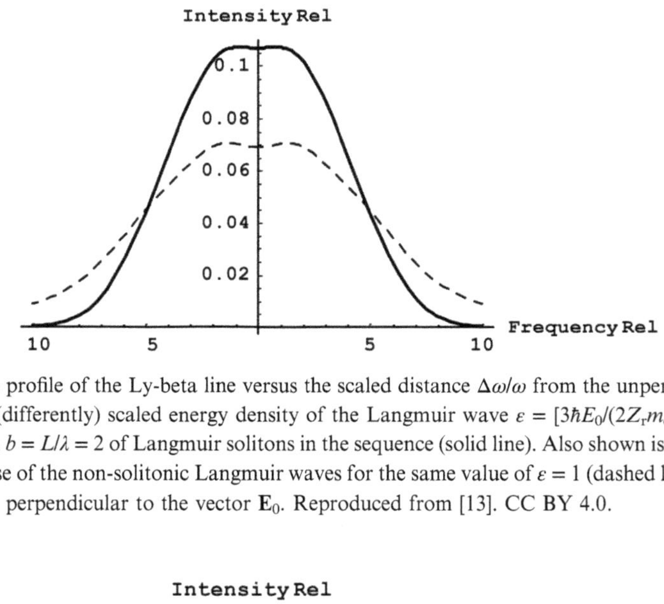

Figure 2.50. The profile of the Ly-beta line versus the scaled distance $\Delta\omega/\omega$ from the unperturbed position of this line for the (differently) scaled energy density of the Langmuir wave $\varepsilon = [3\hbar E_0/(2Z_r m_e e\omega)]^2 = 1$ and the scaled separation $b = L/\lambda = 2$ of Langmuir solitons in the sequence (solid line). Also shown is the corresponding profile for the case of the non-solitonic Langmuir waves for the same value of $\varepsilon = 1$ (dashed line). The direction of observation is perpendicular to the vector $\mathbf{E}_0$. Reproduced from [13]. CC BY 4.0.

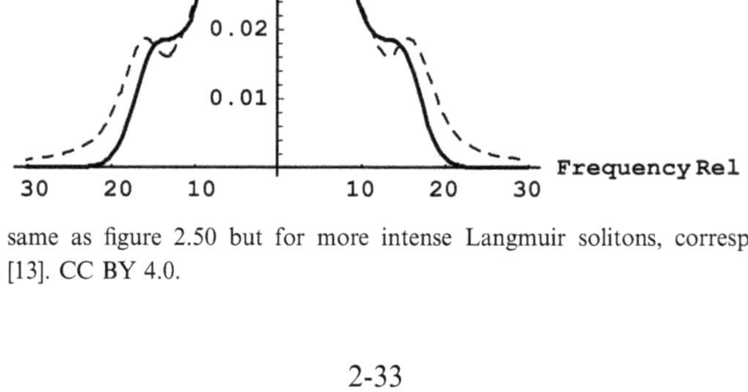

Figure 2.51. The same as figure 2.50 but for more intense Langmuir solitons, corresponding to $\varepsilon = 9$. Reproduced from [13]. CC BY 4.0.

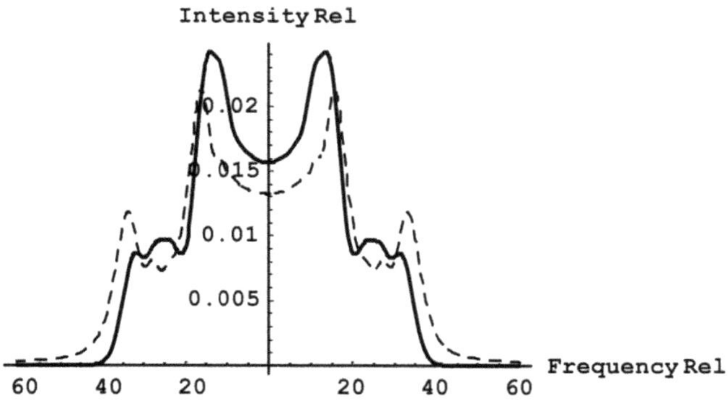

Figure 2.52. The same as figure 2.51 but for even stronger Langmuir waves, corresponding to $\varepsilon = 36$. Reproduced from [13]. CC BY 4.0.

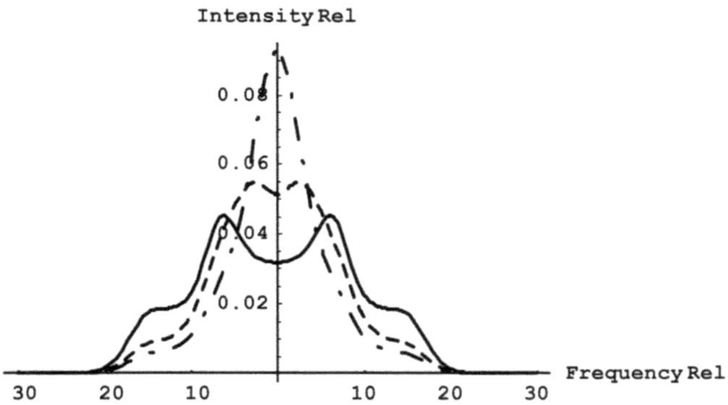

Figure 2.53. The dependence of the Ly-beta profiles for the case of Langmuir solitons, corresponding to the scaled energy density of the solitons $\varepsilon = 9$, on the scaled distance $b = L/\lambda$ between Langmuir solitons in the sequence: $b = 2$ (solid line), $b = 4$ (dashed line), and $b = 6$ (dash-dotted line). Reproduced from [13]. CC BY 4.0.

Paper [13] then discussed the dependence of the solitonic Ly-beta profiles on the scaled separation $b = L/\lambda$ between Langmuir solitons in the sequence. Figure 2.53 presents the Ly-beta profiles for the case of $\varepsilon = 9$ for the following three values of the parameter b: $b = 2$, $b = 4$, and $b = 6$. One can see that with an increase in the scaled separation $b = L/\lambda$ between Langmuir solitons, various features (minima, maxima, and shoulders) gradually fade away.

Thus, the diagnostic of Langmuir solitons based on the use of experimental profiles of the Ly-beta line can be implemented as follows: pronounced secondary maxima would appear in the wings of the spectral line in the non-solitonic situation, but those maxima would appear like shoulders in the solitonic situation—see figures 2.51 and 2.52 and the solid line in figure 2.53. All these transformations of the line profiles are due to their nonlinear dependence on the scaled energy density of the Langmuir solitons.

Bibliography

[1] Blochinzew D I 1933 *Phys. Z. Sow. Union* **4** 501–15

[2] Oks E 2021 *Plasma* **4** 65

[3] Oks E 1995 *Plasma Spectroscopy: The Influence of Microwave and Laser Fields* (Berlin: Springer)

[4] Lifshitz E V 1968 *Sov. Phys. JETP* **26** 570 http://jetp.ras.ru/cgi-bin/e/index/e/26/3/p570?a=list

[5] Zel'dovich J B 1967 *Sov. Phys. JETP* **24** 1006 http://jetp.ras.ru/cgi-bin/e/index/e/24/5/p1006?a=list

[6] Ritus V I 1967 *Sov. Phys. JETP* **24** 1041 http://jetp.ras.ru/cgi-bin/e/index/e/24/5/p1041?a=list

[7] Oks E 2021 *Eur. Phys. J.* D **75** 185

[8] Bernstein I B, Greene J M and Kruskal M D 1957 *Phys. Rev.* **108** 546

[9] Oks E 2017 *Diagnostics of Laboratory and Astrophysical Plasmas Using Spectral Lineshapes of One-, Two-, and Three-Electron Systems* (Singapore: World Scientific)

[10] Oks E and Gavrilenko V P 1983 *Opt. Commun.* **46** 205

[11] Volod'ko D A, Gavrilenko V P and Oks E 1987 *18th Int. Conf. on Phenomena in Ionized Gases (Swansea, UK)* W T Williams (Bristol: Adam Hilger) 604

[12] Kadomtsev B B 1982 *Collective Phenomena in Plasma* (Oxford: Pergamon)

[13] Oks E 2019 *Atoms* **7** 25

Chapter 3

Nonlinear effects in the radiation of hydrogenic spectral lines under the broadband electric field of Langmuir turbulence

Langmuir waves, which are quasimonochromatic in plasmas of relatively low electron densities N_e, can be broadband waves (turbulent) at high electron densities, such as $N_e \gg 10^{18}$ cm^{-3}. The broadening of the power spectrum of Langmuir waves at high electron densities is mainly caused by electron collisions. The corresponding effect on the spectral lines was utilized in papers [1–3] to interpret some of the experimental results. It was also studied theoretically by Gavrilenko [4].

Gavrilenko [4] considered the power spectrum of the stochastic electric field (such as broadband Langmuir turbulence) to be *Lorentzian*. In [5], his results were further developed for the *Gaussian* power spectrum of the stochastic electric field. Also, in distinction to [4], where only hydrogen lines were analyzed, the analysis of [5] was extended to hydrogenic spectral lines; in other words, it included the spectral lines of hydrogenlike ions of any nuclear charge Z. In addition, a new diagnostic method was developed in [5] to measure the Langmuir turbulence average field and to get some information on the Langmuir turbulence power spectrum. We present some details from paper [5] below.

Gavrilenko [4] considered a linearly polarized electric field

$$\mathbf{E}(t) = \sum_{j=1}^{N} \mathbf{E}_j \cos\left(\omega_j t + \varphi_j\right), \quad N \gg 1. \tag{3.1}$$

where $f\mathbf{E}(t)$ was a stationary Gaussian random process characterized by a zero average $\{\mathbf{E}(t)\}_{av} = 0$ and whose correlation function was

$$\{\mathbf{E}(t)\mathbf{E}(t+\tau)\}_{av} = \mathrm{B}G(\tau), \quad \mathrm{B} = \{\mathbf{E}^2\}_{av}. \tag{3.2}$$

In equation (3.2), $G(\tau)$ is a correlation coefficient.

A hydrogenic spectral line, representing the radiative transition between Stark sublevels α of the upper state a and Stark sublevels β of the lower state b, has the following profile $S(\Delta\omega)$:

$$S(\Delta\Omega) = (1/\pi)\sum_{\alpha,\beta} \mathrm{Re} \int_0^\infty \exp(-i\tau\Delta\Omega)\Phi_{\alpha\beta}(\tau)d\tau. \tag{3.3}$$

In equation (3.3), the argument $\Delta\Omega$ is meant to be in the frequency scale and

$$\Phi_{\alpha\beta}(\tau) = \exp(i\delta_{ab}\tau) \left|\langle\varphi_\alpha \mid \mathbf{re} \mid \varphi_\beta\rangle\right|^2 P_{\alpha\beta}(\tau), \tag{3.4}$$

$$P_{\alpha\beta}(\tau) = \exp[-bI(\tau)], \quad b = C_{\alpha\beta}{}^2\mathrm{B}, \quad I(\tau) = \int_0^\tau (\tau - t)G(t)dt, \quad \tau > 0. \tag{3.5}$$

In equations (3.4) and (3.5), $\mathbf{e}$ is the unit vector of the polarization of the emitted photons, $\mathbf{r}$ is the radius vector of the atomic electron, and $\mathbf{re}$ stands for the scalar product of $\mathbf{r}$ and $\mathbf{e}$. The other notations are as follows: the energy difference between levels a and b is denoted by δ_{ab}; φ_α and φ_β are the parabolic wave functions of the hydrogenic atom/ion (the z-axis, which is the parabolic quantization axis, is along $\mathbf{E}(t)$); the quantity $C_{\alpha\beta}$ in equation (3.5) is the Stark constant

$$C_{\alpha\beta} = 3(n_a q_\alpha - n_b q_\beta)/(2Z), \tag{3.6}$$

where n and q are the principal and electric quantum numbers. (Atomic units have been used in equations (3.5) and (3.6): $\hbar = e = m_e = 1$.)

It should be emphasized that the quantity $b = C_{\alpha\beta}{}^2B = C_{\alpha\beta}{}^2\{\mathbf{E}^2\}_{\mathrm{av}}$ is the *scaled energy density* of the Langmuir turbulence. Therefore, we will pay special attention to the dependences of various physical quantities on the parameter b.

The Lorentzian-shaped power spectrum of the stochastic field $\mathbf{E}(t)$ was characterized by the following correlation coefficient:

$$G(\tau) = \exp(-\gamma|\tau|)\cos\omega\tau. \tag{3.7}$$

The analytical result calculated by Gavrilenko [4] for the integral $I(\tau)$ in equation (3.5) was:

$$\begin{aligned} I(\tau) = \{&\exp(-\gamma\tau)[(\gamma^2 - \omega^2)\cos(\omega\tau) - 2\gamma\omega\sin(\omega\tau)] + \gamma\tau(\gamma^2 + \omega^2) \\ &- (\gamma^2 - \omega^2)\}/(\gamma^2 - \omega^2)^2. \end{aligned} \tag{3.8}$$

On substituting equation (3.7) into equation (3.4), he obtained the result for the correlation function $\Phi_{\alpha\beta}(\tau)$. We remind that the spectral line shape is the Fourier transform of the correlation function.

In [5], the author analyzed the case in which the correlation coefficient is

$$G(\tau) = \exp(-\tau^2/g^2)\cos\omega\tau, \tag{3.9}$$

which represents the *Gaussian* form of the power spectrum of the stochastic field. He calculated the integral $I(\tau)$ in equation (3.5) by employing the correlation coefficient $G(\tau)$ from equation (3.9). As a result, he obtained:

$$I(\tau) = (g^2/2)[\exp(-\tau^2/g^2)\cos h(\omega\tau) - 1] + (\pi^{1/2}g/4)\exp(-g^2\omega^2/4) \{\mathrm{Re}[(2\tau + \mathrm{i}g^2\omega)\ \mathrm{erf}(\tau/g + \mathrm{i}g\omega/2)] + \mathrm{i}g^2\omega\ \mathrm{erf}(\mathrm{i}g\omega/2)\}. \tag{3.10}$$

In equation (3.10), erf(z) is the error function. On substituting equation (3.10) into equation (3.4), we the following analytical expression for the correlation function $\Phi_{\alpha\beta}(\tau)$:

$$\Phi_{\alpha\beta}(\tau) = \exp(\mathrm{i}\delta_{ab}\tau)\left|\langle\varphi_\alpha \mid \mathbf{re} \mid \varphi_\beta\rangle\right|^2 \exp\{-b\{(g^2/2)[\exp(-\tau^2/g^2)\cos h(\omega\tau) - 1] + (\pi^{1/2}g/4)\exp(-g^2\omega^2/4) \{\mathrm{Re}[(2\tau + \mathrm{i}g^2\omega)\mathrm{erf}(\tau/g + \mathrm{i}g\omega/2)] + \mathrm{i}g^2\omega\ \mathrm{erf}(\mathrm{i}g\omega/2)\}\}\}. \tag{3.11}$$

It can be seen that in this case, the line profile in equation (3.3) is also the one-fold integral—just as it was in [4].

The author of [5] then considered different limits of the general result of equation (3.11). The first limit was the case in which the carrier frequency is much greater than the characteristic width g of the power spectrum of the stochastic field $\mathbf{E}(t)$:

$$\omega g \gg 1. \tag{3.12}$$

In this situation, equation (3.10) simplifies to

$$I(\tau) = [1 - \cos(\omega\tau)]/\omega^2, \tag{3.13}$$

so that

$$P_{\alpha\beta}(\tau) = \exp\{-b[1 - \cos(\omega\tau)]/\omega^2\}. \tag{3.14}$$

Equation (3.14) for $P_{\alpha\beta}(\tau)$ describes the shape of the spectral line found by Lifshitz [6].

The second limit analyzed in [5]—the more interesting limit physically—was the case in which the carrier frequency is much smaller than the characteristic width g of the power spectrum of the stochastic field $\mathbf{E}(t)$:

$$\omega g \ll 1. \tag{3.15}$$

In this situation, equation (3.10) simplifies to

$$I(\tau) = (g^2/2)[\exp(-\tau^2/g^2) - 1] + (\pi^{1/2}/2)g\tau\ \mathrm{erf}(\tau/g), \tag{3.16}$$

so that

$$P_{\alpha\beta}(\tau) = \exp\{-(bg^2/2)[\exp(-\tau^2/g^2) - 1] + (\pi^{1/2}/2)\mathrm{b}g\tau\ \mathrm{erf}(\tau/g)\}. \tag{3.17}$$

When the time τ is relatively large, so that $\tau/g \gg 1$, the following expression for $P_{\alpha\beta}(\tau)$ was obtained in [5]:

$$P_{\alpha\beta}(\tau) = \exp[-(\pi^{1/2}/2)\mathrm{b}g\tau]. \tag{3.18}$$

This signifies that at relatively small detunings from the unperturbed frequency, the shape of the spectral line is *Lorentzian*. This Lorentzian would have the following full width at half maximum (FWHM):

$$\mathrm{FWHM_L} = \pi^{1/2}\mathrm{bg}. \tag{3.19}$$

When the time τ is relatively small, so that $\tau/g \ll 1$, the author of [5] obtained the following for $P_{\alpha\beta}(\tau)$:

$$P_{\alpha\beta}(\tau) = \exp(-b\tau^2/2). \tag{3.20}$$

This signifies that at relatively large detunings from the unperturbed frequency, the shape of the spectral line changes to Gaussian. This Gasussian shape would have the following FWHM:

$$\mathrm{FWHM_G} = 2(2b\ \ln 2)^{1/2}. \tag{3.21}$$

The author of [5] then let D denote the detuning in the frequency scale from the unperturbed frequency of the spectral line:

$$D = \Delta\Omega - \delta_{ab}. \tag{3.22}$$

For the profile of any Stark component of the hydrogenic spectral line, he then obtained the following (for the situation in which $\omega g \ll 1$):

$$S(D) = (1/\pi)\int_0^\infty \cos(D\tau)\exp\{-(bg^2/2)[\exp(-\tau^2/g^2) - 1] + (\pi^{1/2}/2)\mathrm{bg}\tau\ \mathrm{erf}(\tau/g)\}\mathrm{d}\tau. \tag{3.23}$$

In all figures below, the quantities b, g, and D are measured in units of the carrier frequency ω (for example, $D = (\Delta\Omega - \delta_{\mathrm{ab}})/\omega$). Figure 3.1 presents the calculated profile of any Stark component for $b = 25$ and $g = 0.25$. (Here and below, in all

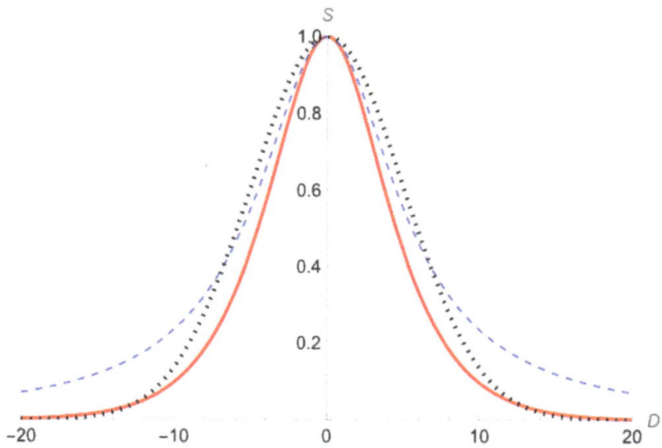

Figure 3.1. The calculated profile of any Stark component for $b = 25$ and $g = 0.25$ (solid line). Also shown is the Lorentzian of $\mathrm{FWHM_L} = \pi^{1/2}bg$ (dashed line) as well as the Gaussian of $\mathrm{FWHM_G} = 2(2b\ \ln 2)^{1/2}$ (dotted line).

figures of this chapter, the quantities b, g, and D are measured in units of the carrier frequency ω: for instance, $D = (\Delta\Omega - \delta_{ab})/\omega$.) Figure 3.1 also displays the Gaussian of $\mathrm{FWHM_G} = 2(2b \ln 2)^{1/2}$ and the Lorentzian of $\mathrm{FWHM_L} = \pi^{1/2}bg$. One can see that the central part of the profile is close to a Lorentzian shape, but in the wings, the shape becomes Gaussian, thus confirming the above analytical results.

To make the transition from the Lorentzian in the profile center to the Gaussian in the wings more visible, figure 3.2 presents parts of the profiles from figure 3.1 using a log[S] versus $D^{1/2}$ representation. One can see that in the wings, the calculated profile (bold line) becomes a straight line in accordance with its transition to the Gaussian.

Figure 3.3 shows how the profile of any Stark component of the spectral line transforms at a fixed value of $b = 40$ as the quantity g varies. One can see that as g

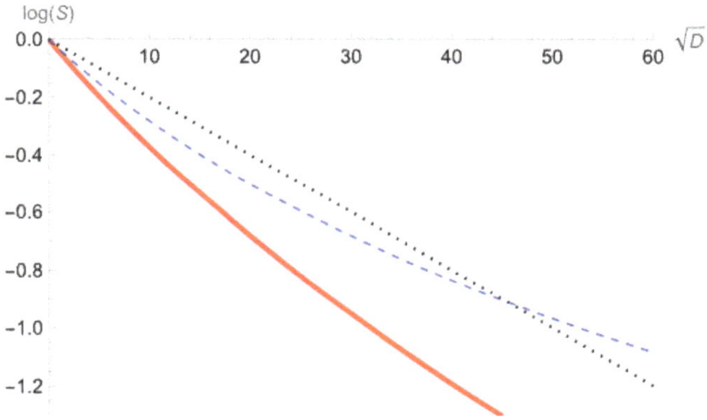

Figure 3.2. Parts of the spectral line profile from figure 3.1 but in a log[S] versus $D^{1/2}$ representation.

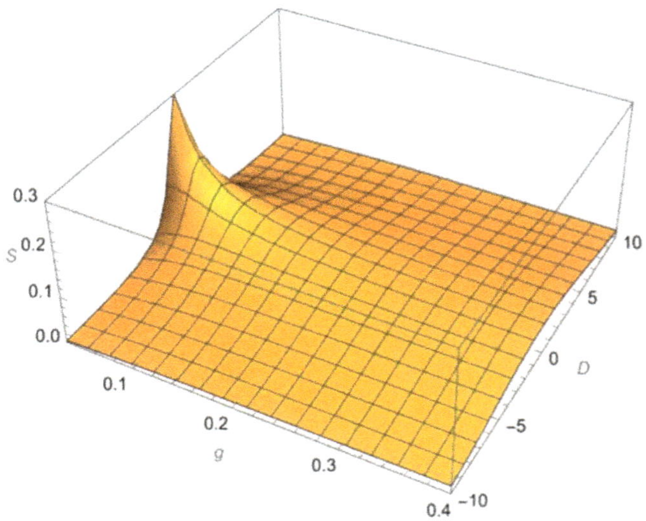

Figure 3.3. The transformation of the profile of any Stark component of the spectral line at a fixed value of $b = 40$ as the quantity g varies.

grows, the width of the spectral profile increases—in accordance with the analytical results presented above.

Figure 3.4 demonstrates the transformation of the profile of any Stark component of the spectral line as the scaled energy density b of the Langmuir turbulence varies at a fixed value of $g = 0.3$. One can see that as the scaled energy density b of the Langmuir turbulence grows, the width of the spectral profile increases—in accordance with the analytical results presented above. If the profile were Lorentzian, then its width would have had a linear dependence on b—according to equation (3.19). The fact that the profile is not actually Lorentzian means that the dependence of the width on the scaled energy density b of the Langmuir turbulence is nonlinear.

Figure 3.5 shows the calculated Ly-beta spectral profiles versus the scaled detuning $D = \Delta\Omega/\omega$ at a fixed value of the scaled root-mean-squared field $\{E^2\}_{av}^{1/2}/(Z_r\omega) = 0.2$ for the following two values of the parameter g that controls the width of the power spectrum of the Langmuir turbulence: $g = 1/4$ and $g = 1/2$. One can see that as the parameter g grows, the spectral line width increases.

It is instructive to compare the profiles of the hydrogenic Ly-beta line for the Gaussian power spectrum utilized in [5] and for the Lorentzian power spectrum employed in [4] for the situation in which both types of power spectrum have the same FWHM. This comparison is presented in figures 3.6 and 3.7. The solid line represents the case of the Gaussian power spectrum and the dashed line represents the case of the Lorentzian power spectrum. Figure 3.6 corresponds to $\{E^2\}_{av}^{1/2}/(Z_r\omega) = 0.2$ and $g = 0.5$. Figure 3.7 corresponds to the $\{E^2\}_{av}^{1/2}/(Z_r\omega) = 5$ and $g = 1$.

It can be seen that that the profile corresponding to the Gaussian power spectrum is narrower than the profile corresponding to the Lorentzian power spectrum. It can also be seen that the difference in the widths of the two profiles increases with the growth of the parameters $\{E^2\}_{av}^{1/2}/(Z_r\omega)$ and g. It is important to emphasize the

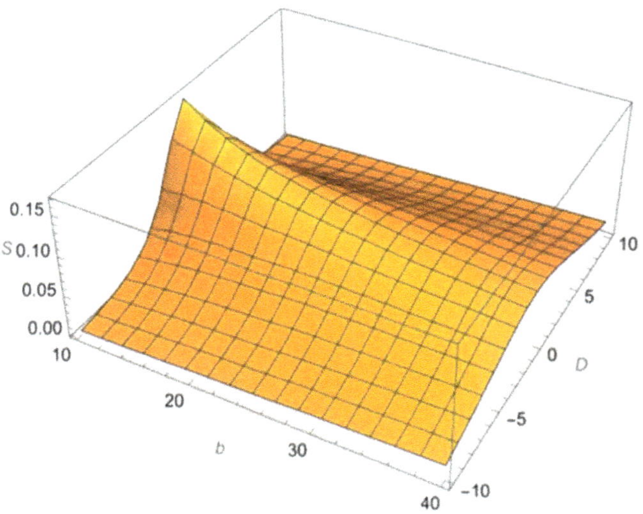

Figure 3.4. The transformation of the profile of any Stark component of the spectral line as the scaled energy density b of the Langmuir turbulence varies at a fixed value of $g = 0.3$.

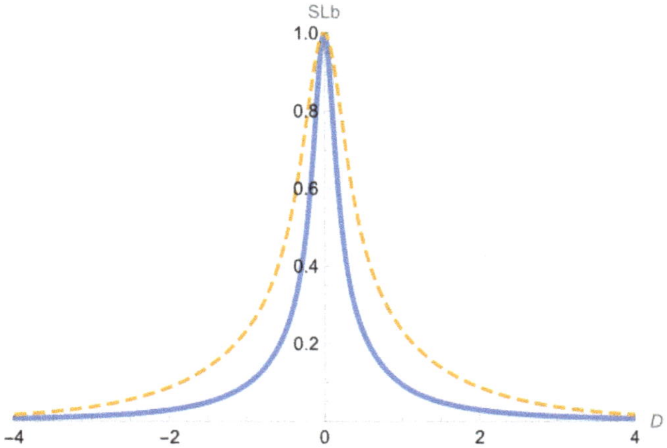

Figure 3.5. The calculated Ly-beta spectral profiles versus the scaled detuning $D = \Delta\Omega/\omega$ at a fixed value of the scaled root-mean-squared field $\{E^2\}_{av}{}^{1/2}/(Z_r\omega) = 0.2$ for the following two values of the parameter g that controls the width of the power spectrum of the Langmuir turbulence: $g = 1/4$ (solid line) and $g = 1/2$ (dashed line).

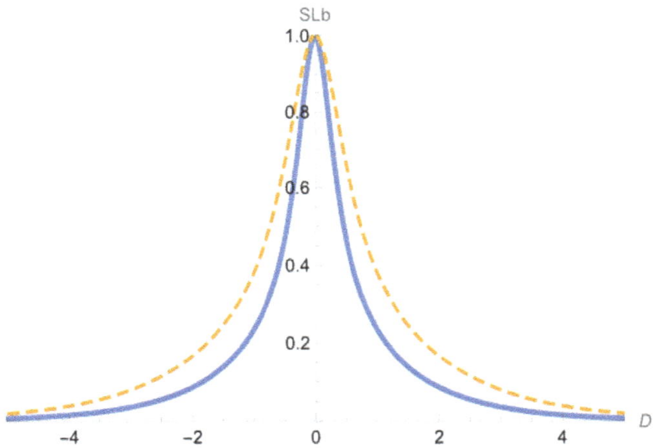

Figure 3.6. A comparison of the profiles of the hydrogenic Ly-beta line for the Gaussian power spectrum utilized in [5] (solid line) and for the Lorentzian power spectrum employed in [4] (dashed line) for the parameters $\{E^2\}_{av}{}^{1/2}/(Z_r\omega) = 0.2$ and $g = 0.5$.

following: at the relatively small value of the scaled field of $\{E^2\}_{av}{}^{1/2}/(Z_r\omega) = 0.2$ and thus a relatively small energy density of the field (figure 3.6), the shapes of both profiles are Lorentzian. However, at the relatively large value of the scaled field of $\{E^2\}_{av}{}^{1/2}/(Z_r\omega) = 5$ and thus a relatively large energy density of the field (figure 3.7), the shapes of both profiles become Gaussian. This effect is nonlinear with respect to the energy density of the stochastic Langmuir turbulence. Figure 3.8 is the same as figure 3.7 but for a smaller value of $g = 0.1$. One can see that for the relatively small g, both profiles are practically the same. Both profiles are Gaussian.

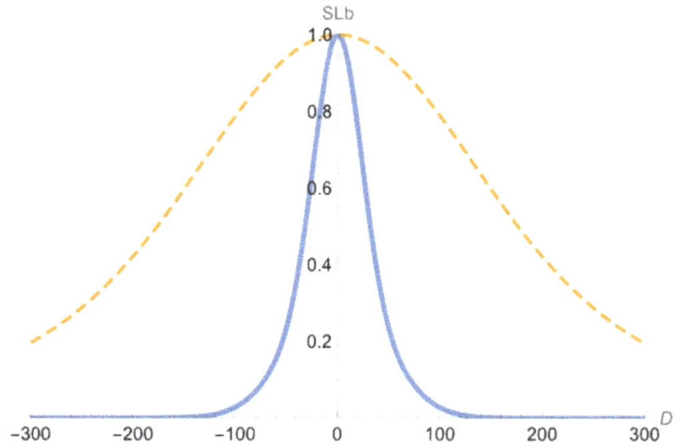

Figure 3.7. The same plot as in figure 3.6 but for the parameters $\{E^2\}_{av}{}^{1/2}/(Z_r\omega) = 5$ and $g = 1$.

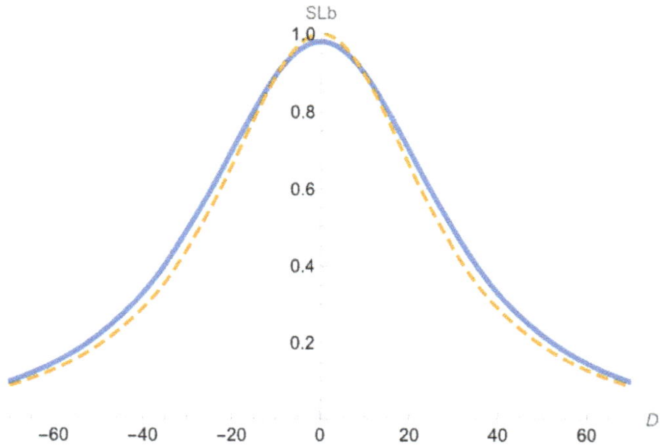

Figure 3.8. The same plot as figure 3.7 but for a smaller value of $g = 0.1$.

Furthermore, [5] revealed that there is a critical value of g: $g_{crit} \sim 0.3$. While at $g < g_{crit}$, the two profiles are almost identical, at $g > g_{crit}$, the two profiles have very different widths: the profile corresponding to the Gaussian power spectrum is narrower than the profile corresponding to the Lorentzian power spectrum.

Finally, [5] emphasized that the obtained results present for the first time the opportunity to deduce from the experimental spectral line profiles *information about the power spectrum* of the stochastic Langmuir turbulence—in addition to measuring the average stochastic field and thus the energy density of the Langmuir turbulence. It also provided a detailed algorithm showing how to do this.

Bibliography

[1] Oks E 2016 *J. Phys. B: At. Mol. Opt. Phys.* **49** 065701

[2] Zakatov L P, Plakhov A G, Shapkin V V and Sholin G V 1971 *Sov. Phys. Dokl.* **16** 451

[3] Karfidov D M and Lukina N A 1997 *Phys. Lett.* A **232** 443
[4] Gavrilenko V P 1996 *Pis'ma v Zh. Tech. Phys. (Sov. Phys. Tech. Phys. Lett.)* **22** 23 (in Russian)
[5] Oks E 2020 *Spectrochim. Acta* B **167** 105815
[6] Lifshitz E V 1958 *Sov. Phys. JETP* **26** 570 http://jetp.ras.ru/cgi-bin/e/index/e/26/3/p570?a=list

Chapter 4

Nonlinear effects in the radiation of satellites of non-hydrogenic spectral lines from plasmas and their applications

4.1 Early theoretical results

This kind of satellite can occur in the spectral lines of helium and lithium atoms, as well as in the spectral lines of He-like and Li-like ions. For example, Baranger and Mozer [1] considered the situation where, in the energy spectrum of a helium atom, one can select a system of three levels, namely zero, one, and two, which have the following properties. Level two is coupled by dipole matrix elements with a closely located level one and to the distant level zero, while levels one and zero are not coupled by a dipole matrix element. So, in the absence of an external electric field, the radiative transition from two to zero is allowed, while the radiative transition from one to zero is forbidden (in the dipole approximation). Under a static electric field **F**, in addition to the allowed spectral line at the frequency $\omega_{20} = \omega_2 - \omega_0$, the dipole-forbidden spectral line appears at the frequency $\omega_{10} = \omega_1 - \omega_0$. (We use atomic units, so that the frequencies ω_0, ω_1, and ω_2 are the energies of the levels zero, one, and two, respectively.)

Under a relatively weak quasimonochromatic electric field (QEF) of frequency ω, two satellites can appear at the frequencies $\omega_{sat} = \omega_{10} \pm \omega$, instead of the dipole-forbidden spectral line. For the case of the isotropic multimode QEF, Baranger and Mozer [1] used the standard nonstationary perturbation theory to obtain the following ratio of the satellite intensities S_+ and S_- to the intensity I_a of the allowed spectral line (in atomic units):

$$S_{\pm}/I_a = [6(\Delta \pm \omega)^2(2l_2 + 1)]^{-1}\langle E^2\rangle \max(l_1, l_2)\left[\int_0^\infty R_{l_1}(r)R_{l_2}(r)r^3 \mathrm{d}r\right]^2, \qquad (4.1)$$

In equation (4.1), l_1 and l_2 are the quantum numbers of the orbital momenta for energy levels one and two; Δ is the separation between these two energy levels; R_l is the radial part of the wave functions in the spherical quantization; the mean-square amplitude of the QEF is denoted by $\langle E^2 \rangle$.

Several years later, Cooper and Ringler [2] performed similar calculations in frames of the standard nonstationary perturbation theory for the case in which the QEF has a single mode and is linearly polarized, being in the form of $\mathbf{E}(t) = \mathbf{E}_0 \cos \omega t$. For observation perpendicular to the field, they obtained the following result:

$$\frac{S_\pm}{I_\mathrm{a}} = \frac{E_0^2}{4(\Delta \pm \omega)^2} \frac{\sum\limits_{m_0, m_1, m_2} |z_{12}|^2 \left(|y_{20}|^2 + |z_{20}|^2\right)}{\sum\limits_{m_0, m_2} \left(|y_{20}|^2 + |z_{20}|^2\right)}. \tag{4.2}$$

The summations in equation (4.2) are performed over the magnetic quantum numbers m_0, m_1, and m_2 of the sublevels of the levels zero, one, and two.

Baranger and Mozer suggested measuring the QEF amplitude by measuring the experimental value of the ratio $S_\pm/I_\mathrm{a}$ and then comparing it with the corresponding theoretical ratio. This idea was first implemented experimentally by Kunze and Griem [3] and was then employed by other authors in several experiments. However, these experiments soon entered the situation in which the QEF was relatively strong, so that the Baranger–Mozer theoretical results [1] and the Cooper–Ringler theoretical results [2] became invalid.

4.2 The adiabatic theory of satellites

The overwhelming majority of these later experiments corresponded to the low-frequency case, where $\omega \ll \Delta$. In this case, the employment of the adiabatic perturbation theory allowed the corresponding analytical results to be obtained beyond the limits of validity of the standard nonstationary perturbation theory—see section 5.1 of [3]. To illustrate this, we let:

$$\alpha = 2E_0 z_{12}/\Delta. \tag{4.3}$$

The nonstationary perturbation theory is valid only for $\alpha \ll 1$. In contrast, the adiabatic perturbation theory is valid for

$$\alpha\omega/\Delta \ll 1. \tag{4.4}$$

So, as long as $\omega \ll \Delta$, condition (4.4) can be satisfied even for $\alpha > 1$, that is, for the stronger QEF. The analytical results obtained using the adiabatic perturbation theory (the results of which are presented below) allowed the experimental determination of the QEF amplitude for relatively strong fields.

The adiabatic perturbation theory of satellites starts from the instantaneous eigenvalues of the operator

$$\mathrm{H}(t) = \mathrm{H}_\mathrm{a} + zE_0 \cos \omega t \tag{4.5}$$

(where H_a is the Hamiltonian of the unperturbed atom) for the two-level subsystem consisting of levels one and two:

$$\omega_{1,2}(t) = [\omega_1^{(0)} + \omega_2^{(0)} \pm (-\Delta)(1 + \alpha^2 \cos^2 \omega t)^{1/2}]/2. \tag{4.6}$$

The corresponding instantaneous eigenfunctions that constitute the adiabatic basis, are

$$\begin{aligned} \chi_1(t) &= \psi_1 \cos(\beta/2) - \psi_2 \sin(\beta/2), \\ \chi_2(t) &= \psi_1 \cos(\beta/2) - \psi_2 \sin(\beta/2), \\ \beta &\equiv \arctan(\alpha \cos \omega t). \end{aligned} \tag{4.7}$$

The solution of the Schrödinger equation

$$\mathrm{i}\, \partial\psi/\partial t = (\mathrm{H}_a + zE_0 \cos \omega t)\psi \tag{4.8}$$

is sought in the form

$$\psi(t) = \sum_{j=1}^{2} C_j(t)\chi_j(t) exp\left[-\mathrm{i}\int_0^t \mathrm{d}t'\omega_j(t')\right]. \tag{4.9}$$

On substituting equation (4.9) into equation (4.8), one obtains:

$$\begin{aligned} \dot{C}_1 &= -\, C_2(\dot{\beta}/2)exp\left[-\mathrm{i}\int_0^t \mathrm{d}t'\omega_{21}(t')\right] \\ \dot{C}_2 &= C_1(\dot{\beta}/2)exp\left[\mathrm{i}\int_0^t \mathrm{d}t'\omega_{21}(t')\right]; \end{aligned} \tag{4.10}$$

$$\dot{\beta} = -\alpha\omega(sin\omega t)/(1 + \alpha^2 cos^2\omega t). \tag{4.11}$$

To calculate the intensities of the satellites, the following initial conditions are chosen: $C_1(0) = 1$, $C_2(0) = 0$. With these initial conditions, the solution of the system of equations (4.10) is

$$C_2(t) \approx \frac{1}{2}\int_0^t \mathrm{d}t'\dot{\beta}(t')\exp\left(\mathrm{i}\int_0^{t'} \mathrm{d}\tau\omega_{21}(t)\right). \tag{4.12}$$

In order to calculate the integrals in equation (4.12), the quantities $\dot{\beta}(t)$ and $\omega_{21}(t)$ have been expanded into the corresponding Fourier series:

$$\begin{aligned} \dot{\beta}(t) &= \mathrm{i}(k\omega/4) \sum_{p=-\infty}^{+\infty} (a_{2p} - a_{2p+2})exp[i(2p + 1)\omega t], \\ \omega_{21}(t) &= \overline{\Delta} + \sum_{q=1}^{\infty} \varepsilon_{2q} \cos 2q\omega, \\ \overline{\Delta} &\equiv (2/\pi)\Delta(1 + \alpha^2)^{1/2} E(k). \end{aligned} \tag{4.13}$$

The notations used in equation (4.13) are as follows:

$$k \equiv \alpha(1 + \alpha)^{-1/2}, \tag{4.14}$$

$$a_{2p} = 2(-1)^p k^{-2|p|}[(1 - k^2)^{1/2} - 1]^{2|p|}, \tag{4.15}$$

$$\varepsilon_2 = (4/3\pi)\Delta(1 + \alpha^2)^{1/2}[\mathbb{E}(k) - 2(1 - k^2)\mathbb{D}(k)], \tag{4.16}$$

$$\varepsilon_4 = (4(15\pi k^4))^{-1}\Delta \times (1 + \alpha^2)^{1/2}[(8k^4 - 24k^2 + 16)\mathbb{K}(k) + (-k^4 + 16k^2 - 16)\mathbb{E}(k)]. \tag{4.17}$$

In equations (4.16) and (4.17), $E(k)$, $D(k)$, and $K(k)$ are the complete elliptic integrals.

To illustrate the last line of equation (4.13) pictorially in terms of its dependence on the energy density of the field, we rewrite it in the form

$$D = (2/\pi)\Delta(1 + a/\Delta^2)^{1/2} E[(a^{1/2}/\Delta)/(1 + a/\Delta^2)^{1/2}], \tag{4.18}$$

where

$$a = 4z^{1/2}E_0{}^2 \tag{4.19}$$

is the scaled energy density of the field. Figure 4.1 shows the dependence of the perturbed separation D of levels one and two versus the scaled energy density a and the unperturbed separation Δ, all quantities being in atomic units.

Figure 4.2 shows the same results as those of figure 4.1 but from the alternative viewpoint, so that together with figure 4.1, it demonstrates in a comprehensive way the nonlinear dependence of the perturbed separation D of levels one and two on the scaled energy density a of the laser field for various values of the unperturbed separation Δ.

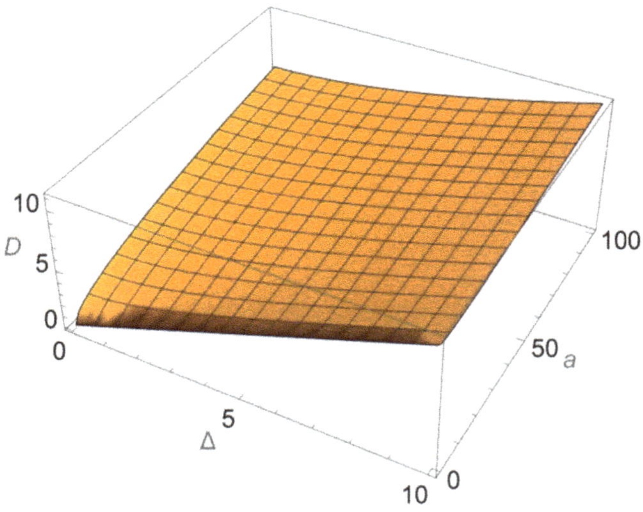

Figure 4.1. The dependence of the perturbed separation D of levels one and two versus the scaled energy density a and the unperturbed separation Δ, all quantities being in atomic units (see equations (4.18) and (4.19)).

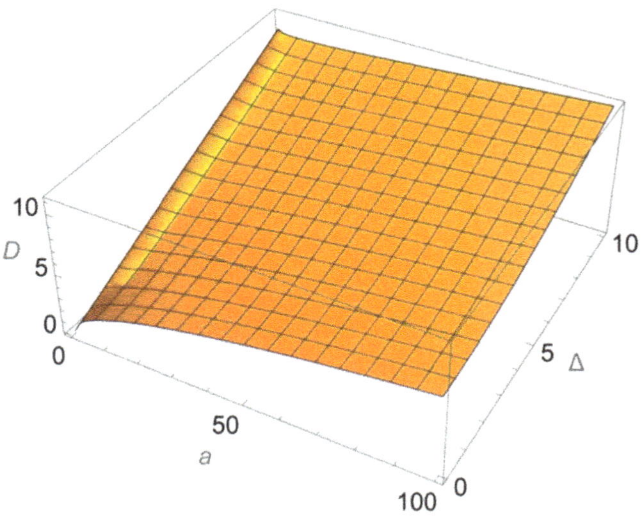

Figure 4.2. The same results as those shown in figure 4.1 but from the alternative viewpoint, so that together with figure 4.1, this figure demonstrates in a comprehensive way the nonlinear dependence of the perturbed separation D of levels one and two on the scaled energy density a of the laser field.

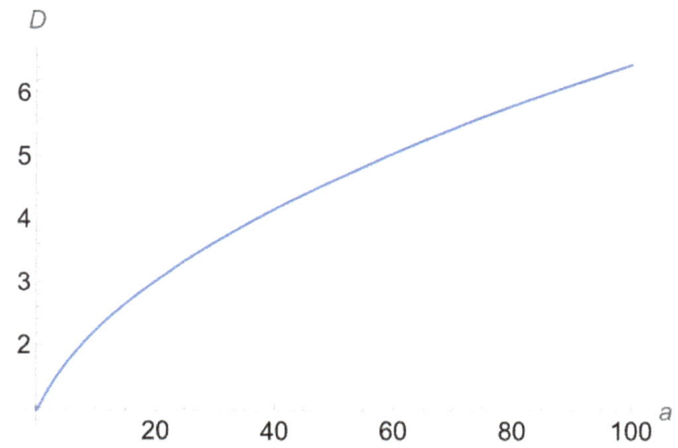

Figure 4.3. The dependence of the perturbed separation D of levels one and two on the scaled energy density a of the laser field at the unperturbed separation $\Delta = 1$.

Figure 4.3 presents the dependence of the perturbed separation D of levels one and two on the scaled energy density a of the laser field at the unperturbed separation $\Delta = 1$.

From figures 4.1–4.3, it can be seen that the dependence of the perturbed separation D of levels one and two on the scaled energy density a of the laser field is significantly nonlinear.

Figure 4.4 shows the dependence of the Fourier expansion coefficient ε_{2q} from equation (4.13) on the scaled energy density a and the unperturbed separation Δ for $q = 1$.

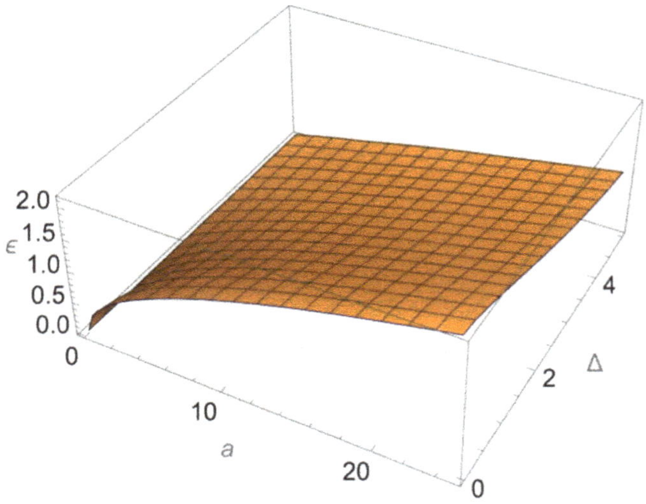

Figure 4.4. The dependence of the Fourier expansion coefficient ε_{2q} from equation (4.13) on the scaled energy density a and the unperturbed separation Δ for $q = 1$.

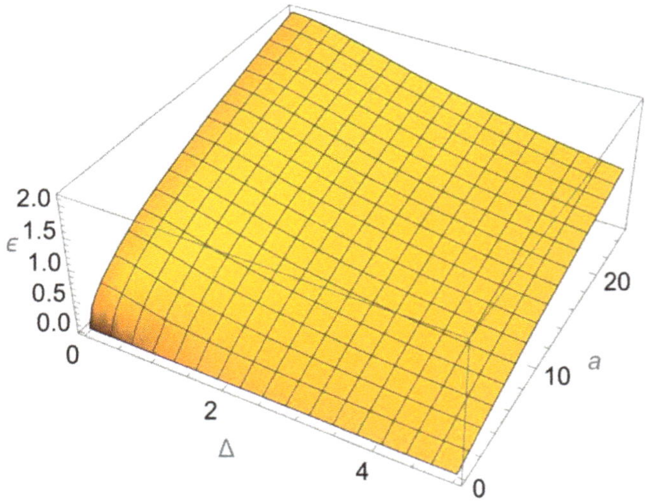

Figure 4.5. The same results as those shown in figure 4.4 but from the alternative viewpoint, so that together with figure 4.4, this figure demonstrates in a comprehensive way the nonlinear dependence of the Fourier expansion coefficient ε_{2q} from equation (4.13) on the scaled energy density a and the unperturbed separation Δ for $q = 1$.

Figure 4.5 shows the same results as those of figure 4.4 but from the alternative viewpoint, so that together with figure 4.4, it demonstrates in a comprehensive way the nonlinear dependence of the Fourier expansion coefficient ε_{2q} from equation (4.13) on the scaled energy density a and the unperturbed separation Δ for $q = 1$.

Figure 4.6 shows the dependence of the Fourier expansion coefficient ε_{2q} from equation (4.13) on the scaled energy density a and the unperturbed separation Δ for $q = 2$.

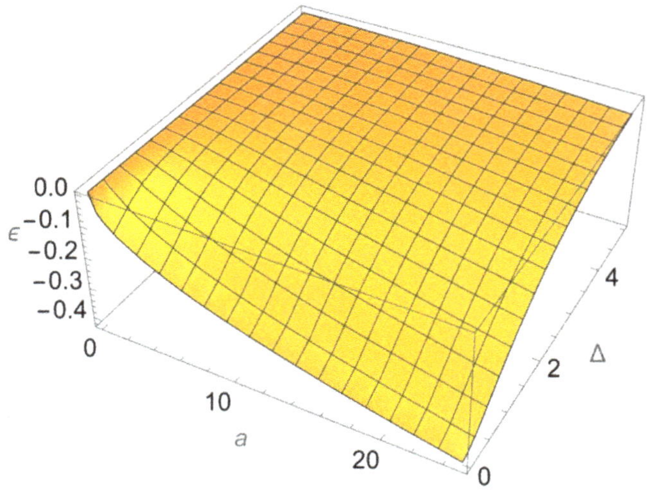

Figure 4.6. The dependence of the Fourier expansion coefficient ε_{2q} from equation (4.13) on the scaled energy density a and the unperturbed separation Δ for $q = 2$.

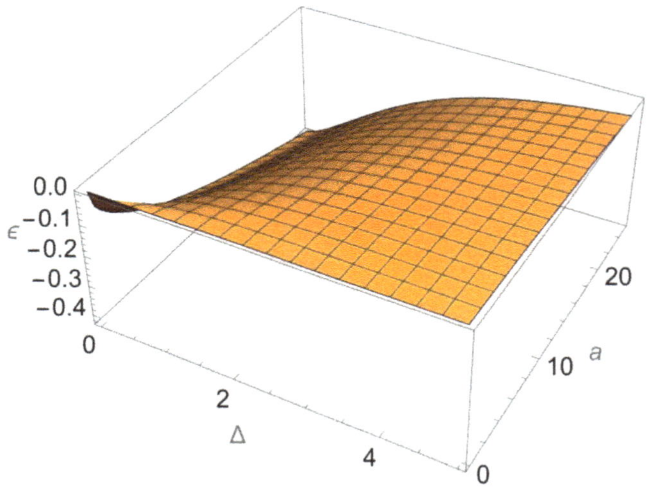

Figure 4.7. The same results as those of figure 4.6 but from the alternative viewpoint, so that together with figure 4.6, this figure demonstrates in a comprehensive way the nonlinear dependence of the Fourier expansion coefficient ε_{2q} from equation (4.13) on the scaled energy density a and the unperturbed separation Δ for $q = 2$.

Figure 4.7 shows the same results as those of figure 4.6 but from the alternative viewpoint, so that together with figure 4.6, it demonstrates in a comprehensive way the nonlinear dependence of the Fourier expansion coefficient ε_{2q} from equation (4.13) on the scaled energy density a and the unperturbed separation Δ for $q = 2$.

Figure 4.8 shows the dependence of the Fourier expansion coefficient ε_{2q} from equation (4.13) on the scaled energy density a and the unperturbed separation Δ for $q = 3$.

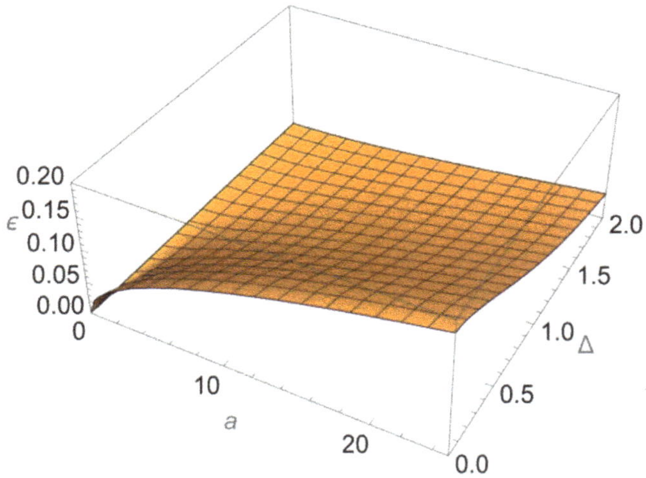

Figure 4.8. The dependence of the Fourier expansion coefficient ε_{2q} from equation (4.13) on the scaled energy density a and the unperturbed separation Δ for $q = 3$.

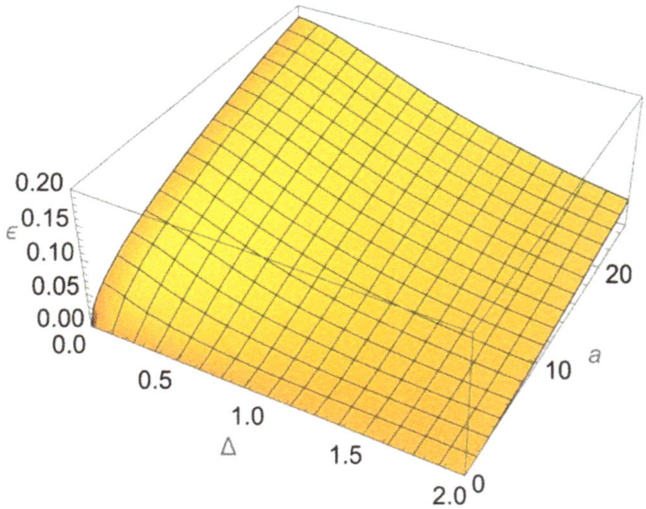

Figure 4.9. The same results as those shown in figure 4.8 but from the alternative viewpoint, so that together with figure 4.8, this figure demonstrates in the comprehensive way the nonlinear dependence of the Fourier expansion coefficient ε_{2q} from equation (4.13) on the scaled energy density a and the unperturbed separation Δ for $q = 3$.

Figure 4.9 shows the same results as those of figure 4.8 but from the alternative viewpoint, so that together with figure 4.8, this figure demonstrates in a comprehensive way the nonlinear dependence of the Fourier expansion coefficient ε_{2q} from equation (4.13) on the scaled energy density a and the unperturbed separation Δ for $q = 3$.

Figure 4.10 shows the dependence of the Fourier expansion coefficient ε_{2q} from equation (4.13) on the scaled energy density a and the unperturbed separation Δ for $q = 4$.

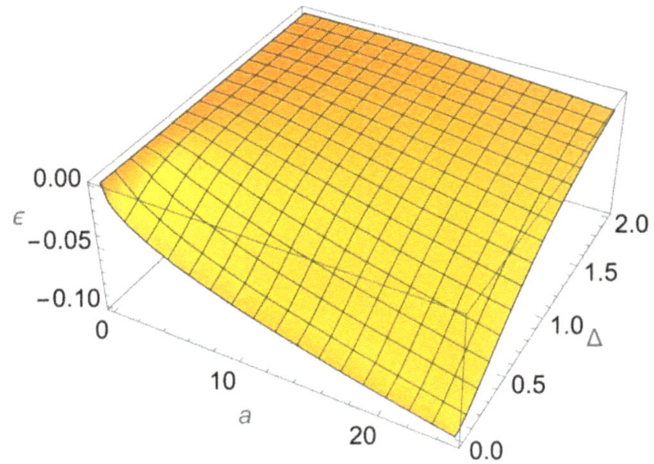

Figure 4.10. The dependence of the Fourier expansion coefficient ε_{2q} from equation (4.13) on the scaled energy density a and the unperturbed separation Δ for $q = 4$.

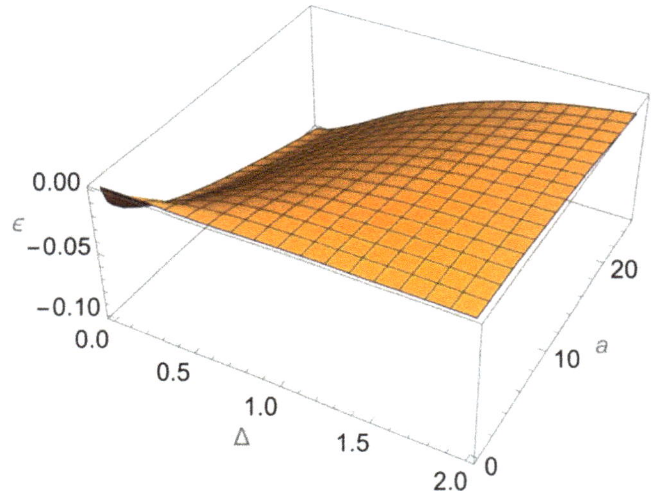

Figure 4.11. The same results as those shown in figure 4.10 but from the alternative viewpoint, so that together with figure 4.8, this figure demonstrates in a comprehensive way the nonlinear dependence of the Fourier expansion coefficient ε_{2q} from equation (4.13) on the scaled energy density a and the unperturbed separation Δ for $q = 4$.

Figure 4.11 shows the same results as those shown in figure 4.10 but from the alternative viewpoint, so that together with figure 4.8, it demonstrates in a comprehensive way the nonlinear dependence of the Fourier expansion coefficient ε_{2q} from equation (4.13) on the scaled energy density a and the unperturbed separation Δ for $q = 4$.

From figures 4.2 and 4.4–4.11, it can be seen that the dependence of the Fourier expansion coefficients ε_{2q} from equation (4.13) on the scaled energy density a is significantly nonlinear for any q.

The solution of equation (4.12) is

$$C_2(t) \approx \begin{aligned}&(k\omega/8)\{[(2-a_2)J_0 - 2J_1]g(\overline{\Delta}+\omega) - [(2-a_2)J_0 + 2J_1]\\ &\times g(\overline{\Delta}-\omega) + (2J_1 + a_2J_0)g(\overline{\Delta}+3\omega) + (2J_1 - a_2J_0)g(\overline{\Delta}-3\omega)\},\end{aligned} \quad (4.20)$$

$$g(u) \equiv [\exp(\mathrm{i}ut) - 1]/u,$$

where J_0 and J_1 stand for the Bessel functions $J_0[\varepsilon_2/(2\omega)]$ and $J_1[\varepsilon_2/(2\omega)]$. The corresponding approximate solution of equation (4.8) is

$$\psi(t) \approx \chi_1(t)exp\left[-\mathrm{i}\int_0^t \mathrm{d}t'\omega_1(t')\right] + C_2(t)\chi_2(t)exp\left[-\mathrm{i}\int_0^t \mathrm{d}t'\omega_2(t')\right]. \quad (4.21)$$

The spectrum of the spontaneous emission to level zero is determined by the following equation

$$I^{(e)}(\Delta\omega) = \lim_{T\to\infty}(2\pi T)^{-1}\left|\int_0^T \mathrm{d}t\langle\psi(t)|re|\psi_0\rangle\exp(-\mathrm{i}t\Delta\omega)\right|^2, \quad (4.22)$$

where the unit vector of photon polarization is denoted by **e**.

In many experiments, the satellites are emitted from a much smaller volume than the allowed spectral line. In this situation, both the theoretical ratio of the intensity S_- of the near satellite to the intensity I_a of the allowed line and the theoretical ratio of the intensity S_+ of the far satellite to the intensity I_a of the allowed line cannot be used to determine the field amplitude E_0. A possible way to resolve this is to compare the experimental ratio of the satellite intensities S_-/S_+ with the corresponding theoretical ratio. However, within the first nonvanishing order of the standard perturbation theory ($\sim E_0^2$), this theoretical ratio does not depend on E_0 and cannot be used to determine E_0. To obtain the dependence of the theoretical ratio S_-/S_+ within the standard perturbation theory, one should extend the calculations to at least the next order ($\sim E_0^4$). The corresponding result for the satellite intensities $\sigma_\pm$ is

$$\sigma_\pm \approx \frac{1}{(\Delta \pm \omega)^2}\left[\left(\frac{E_0 z_{12}}{2}\right)^2 \pm \left(\frac{E_0 z_{12}}{2}\right)^4 \frac{\Delta^3 \mp 7\Delta^2\omega + 3\Delta\omega^2 \mp \omega^3}{\omega(\Delta^2 - \omega^2)^2}\right]. \quad (4.23)$$

Figure 4.12 presents (in solid lines) the dependence of the satellite intensity ratio S_-/S_+ on the field amplitude E_0 for the lines He I 4471 Å and He I 4922 Å, calculated by the adiabatic theory for the frequency $\omega/(2\pi c) = 1.28\ \mathrm{cm}^{-1}$. For comparison, the dashed lines show the corresponding results calculated by the standard perturbation theory up to $\sim E_0^4$.

One can see that even after the inclusion of the terms $\sim E_0^4$, the standard perturbation theory can be used to determine the field amplitude E_0 from the corresponding experimental ratio alone for relatively small fields. In contrast, the adiabatic theory opens up a way to extend this diagnostic method to stronger fields.

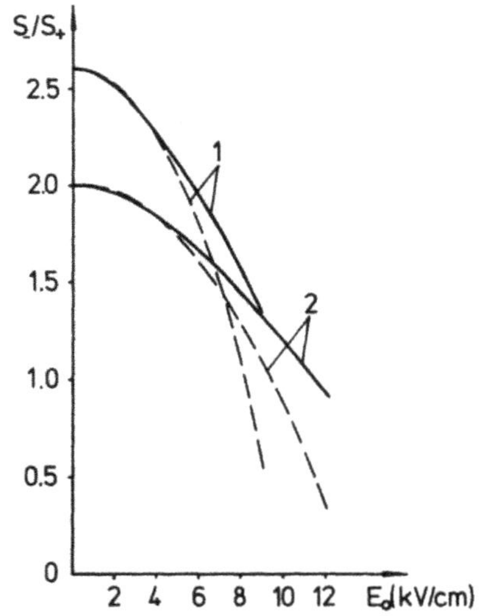

Figure 4.12. Solid lines—the dependence of the satellite intensity ratio S_-/S_+ on the field amplitude E_0 for the lines He I 4471 Å and He I 4922 Å, calculated by the adiabatic theory for the frequency $\omega/(2\pi c) = 1.28\ \text{cm}^{-1}$. Dashed lines—the corresponding results calculated by the standard perturbation theory up to $\sim E_0^4$. The lines marked '1' are for He I 4471 Å and the lines marked '2' are for He I 4922 Å. Reprinted from [3], Copyright (1995), with permission from Springer.

4.3 The high-frequency or strong field case in the three-level scheme

We consider the same system of two closely lying levels one and two and the distant level zero. The effect of the linearly polarized field $\mathbf{E}(t) = \mathbf{E}_0 \cos \omega t$ on the subsystem of levels one and two in the high-frequency case turns out to be equivalent to the effect in the strong field case. This is because the corresponding small parameter is

$$\delta = \Delta^2/(z_{12}E_0\omega) \ll 1. \tag{4.24}$$

The corresponding solution for the probability amplitudes $C_1(t)$ and $C_2(t)$ can be represented in the form [4]

$$\begin{gathered} C_{1,\,2}(t) = 2^{-3/2} \exp\left[-\mathrm{i}(\omega_1^{(0)} + \omega_2^{(0)})t/2\right] \\ \{\cos(Qt)\exp[-(\mathrm{i}A)\sin\omega t] \pm \sin(Qt)\exp[(\mathrm{i}A)\sin\omega t]\}, \end{gathered} \tag{4.25}$$

where

$$Q = \Delta J_0(2A)/2, \qquad A = z_{12}E_0/\omega. \tag{4.26}$$

The quantities $\pm Q$ are the quasienergies. To study the dependence of the quasienergies on the energy density of the field, we introduce the following scaled energy density of the field:

$$B = A^2 = (z_{12}E_0/\omega)^2. \tag{4.27}$$

Figure 4.13 shows a plot of the quasienergy Q versus the scaled energy density of the field and the unperturbed separation Δ of levels one and two. We remind the reader that all quantities are in atomic units.

Figure 4.14 shows the same results as those of figure 4.13 but from the alternative viewpoint, so that together with figure 4.13, it demonstrates in a comprehensive way the nonlinear dependence of the quasienergy Q on the scaled energy density of the field B and on the unperturbed separation Δ of levels one and two.

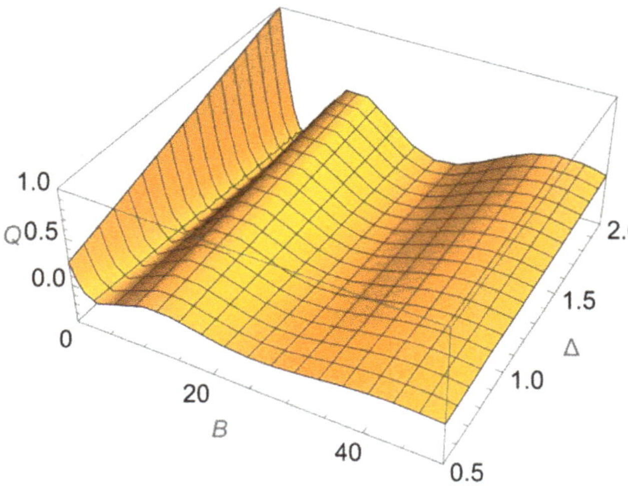

Figure 4.13. The dependence of the quasienergy Q from equation (4.26) on the scaled energy density of the field B defined in equation (4.27) and on the unperturbed separation Δ of the levels one and two.

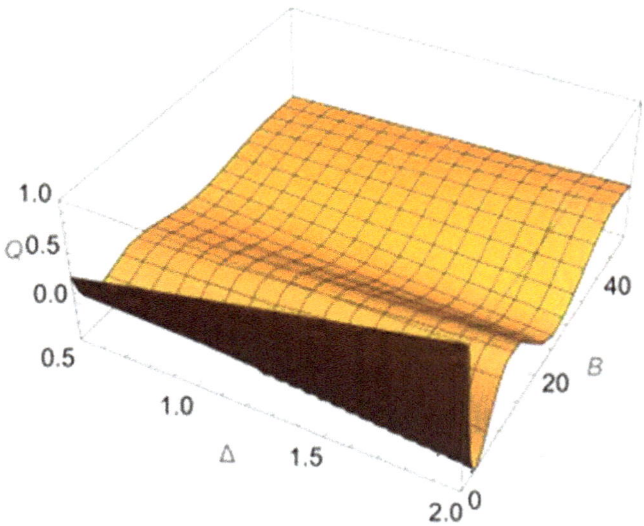

Figure 4.14. The same results as those shown in figure 4.13 but from the alternative viewpoint, so that together with figure 4.13, this figure demonstrates in a comprehensive way the nonlinear dependence of the quasienergy Q on the scaled energy density of the field B and on the unperturbed separation Δ of levels one and two.

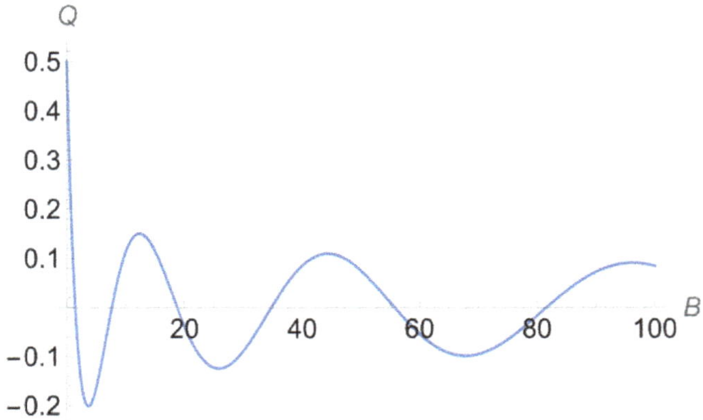

Figure 4.15. The dependence of the quasienergy Q on the scaled energy density of the field B for the unperturbed separation (of levels one and two) $\Delta = 1$.

Figure 4.15 shows the dependence of the quasienergy Q on the scaled energy density of the field B for the unperturbed separation (of levels one and two) $\Delta = 1$.

From figures 4.13–4.15, it can be seen that the dependence of the quasienergy Q on the energy density of the field is highly nonlinear. From equations (4.26) and (4.27), it is clear that as the energy density of the field B increases, the separation $2Q$ of the two quasienergies decreases. In this sense, it can be stated that the mutual 'repulsion' of levels one and two, which is characteristic for relatively small fields ($\delta \gg 1$, $\omega \leqslant \Delta$), changes to mutual 'attraction' for relatively strong fields ($\delta \ll 1$, $\omega \geqslant \Delta$).

From equation (4.25), it can be seen that in the spectrum of the radiative transition from levels one and two to the relatively distant level zero, two systems of satellites appear at the frequencies

$$\omega_{\text{sat},\,1} = Q + p\omega \tag{4.28}$$

and the frequencies

$$\omega_{\text{sat},\,2} = -Q + q\omega, \tag{4.29}$$

where $p = 0, \pm 1, \pm 2, \ldots$ and $q = 0, \pm 1, \pm 2, \ldots$. As the amplitude or the frequency of the field increases, this spectrum approaches Blochinzew's spectrum [5].

4.4 The high-frequency or strong field case in the four-level scheme

We consider a system of three closely lying levels one, two, and three and a distant level zero. The effect of the linearly polarized field $\mathbf{E}(t) = \mathbf{E}_0 \cos \omega t$ on the subsystem of levels one, two, and three in the high-frequency case turns out to be equivalent to the effect in the strong field case—similarly to the situation presented in the previous section. The subsystem of levels one, two, and three is characterized by the following dipole matrix elements: $z_{12} \neq 0$, $z_{13} \neq 0$, $z_{23} = 0$.

The wave functions are sought in the following form

$$\psi(t) = \sum_{n=1}^{3} C_n(t)\varphi_n(t),\ \varphi_n(t) = exp(-i\omega^{-1}E_0\hat{z}\ \sin\omega t)\psi_n^{(0)}, \tag{4.30}$$

where $\psi_n^{(0)}$ are the unperturbed wave functions. We remind the reader again that we utilize the atomic units: $\hbar = m = e = 1$. After substituting equation (4.30) into the corresponding Schrödinger equation, one gets

$$\begin{aligned}
\dot{C}_1 &= -\,iC_1V\sin^2\alpha + C_2\tilde{z}_{12}\sin\alpha(\omega_{23}\tilde{z}_{13}^2 + V\cos\alpha)\\
&\quad + C_3\tilde{z}_{13}\sin\alpha(-\omega_{23}\tilde{z}_{13}^2 + V\cos\alpha),\\
\dot{C}_2 &= -\,C_1\tilde{z}_{12}\sin\alpha(\omega_{23}\tilde{z}_{13}^2 + V\cos\alpha) - iC_2(W\tilde{z}_{13}^2 + 2\tilde{z}_{12}^2\tilde{z}_{13}^2\omega_{23}\cos\alpha\\
&\quad + V\tilde{z}_{12}^2\cos^2\alpha) - iC_3\tilde{z}_{12}\\
&\qquad \tilde{z}_{13}[-W + (\tilde{z}_{13}^2 - \tilde{z}_{12}^2)\omega_{23}cos\alpha + V\cos^2\alpha],
\end{aligned} \tag{4.31}$$

where

$$\begin{aligned}
&\alpha \equiv B\sin\omega t,\ B \equiv \xi E_0/\omega,\ \xi \equiv (z_{12}^2 + z_{13}^2)^{1/2},\\
&\tilde{z}_{12} \equiv z_{12}/\xi,\ \tilde{z}_{13} \equiv z_{13}/\xi,\\
&V \equiv \omega_{21}\tilde{z}_{12}^2 + \omega_{31}\tilde{z}_{13}^2,\ W = \omega_{21}\tilde{z}_{13}^2 + \omega_{31}\tilde{z}_{12}^2.
\end{aligned} \tag{4.32}$$

The corresponding equation for the probability amplitude C_3 can be produced from the equation for the probability amplitude C_2 by transposing the indices $2 \leftrightarrow 3$.

To obtain an approximate solution, we use the following Fourier expansions:

$$\begin{aligned}
\sin\alpha &= 2\sum_{k=1}^{\infty} J_{2k-1}(B)\sin(2k-1)\omega t,\ \cos\alpha = J_0(B)\\
&\quad + 2\sum_{k=1}^{\infty} J_{2k}(B)\sin 2k\omega t
\end{aligned} \tag{4.33}$$

and neglect the oscillatory parts of the coefficients of the system of equation (4.31). As a result, this system of equations takes the following form:

$$\dot{C}_1 = -ia_{11}C_1,\ \dot{C}_2 = -ia_{22}C_2 - ia_{23}C_3,\ \dot{C}_3 = -ia_{32}C_2 - ia_{32}C_3, \tag{4.34}$$

where

$$\begin{aligned}
&a_{11} = [1 - J_0(2B)]V/2;\\
&a_{23} = a_{32} = \tilde{z}_{12}\tilde{z}_{13}\{(\tilde{z}_{13}^2 - \tilde{z}_{12}^2)\omega_{23}[J_0(B) - 1] + [J_0(2B) - 1]V/2\};\\
&a_{pp} = \omega_{p1} + (-1)^p 2\tilde{z}_{12}^2\tilde{z}_{13}^2\omega_{23}[J_0(B) - 1]\\
&\quad + \tilde{z}_{1p}^2[J_0(2B) - 1]V/2,\ (p = 2, 3).
\end{aligned} \tag{4.35}$$

After finding the solution of the system (4.34), we obtain the following wave functions of the quasienergy states in the high-frequency or strong field $\mathbf{E}(t)$:

$$\psi_1(t) = \varphi_1 \exp(-i\mu_1 t),\ \psi_2(t) = (1 + A^2)^{-1/2}(\varphi_2 - A\varphi_3)exp(-i\mu_2 t),$$
$$\psi_3(t) = (1 + A^2)^{-1/2}(A\varphi_2 + \varphi_3)exp(-i\mu_3 t), \tag{4.36}$$
$$A \equiv (\mu_3 - a_{33})/a_{23} = (a_{22} - \mu_2)/a_{23},$$

where

$$\mu_1 = a_{11},\quad \mu_{2,\,3} = (a_{22} + a_{33})/2 \pm [(a_{22} - a_{33})^2/4 + a_{23}^2]^{1/2} \tag{4.37}$$

are the quasienergies.

We introduce the following scaled energy density of the field:

$$G = B^2 = (\xi E_0/\omega)^2. \tag{4.38}$$

As an example, figure 4.16 shows the dependence of the quasienergy μ_1 on the scaled energy density of the field G and on the parameter V defined in equation (4.32).

Figure 4.17 shows the same results as those of figure 4.16 but from the alternative viewpoint, so that together with figure 4.16, it demonstrates in a comprehensive way the nonlinear dependence of the quasienergy μ_1 on the scaled energy density of the field G and on the parameter V.

Figure 4.18 shows the dependence of the quasienergy μ_1 on the scaled energy density of the field G for the parameter $V = 5$.

From figures 4.16–4.18, it can be seen that the dependence of the quasienergy μ_1 on the scaled energy density of the field G is highly nonlinear.

Appendix A presents an example of the application of the above general results to the specific case in which the three levels from which the radiative transitions originate are the energy levels 5D, 5F, and 5G of helium.

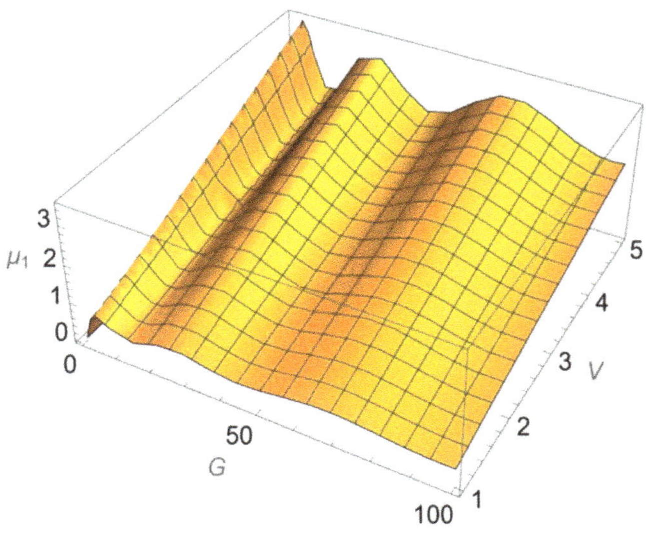

Figure 4.16. The dependence of the quasienergy μ_1 from equation (4.37) (see also equation (4.35)) on the scaled energy density of the field G and on the parameter V defined in equation (4.32).

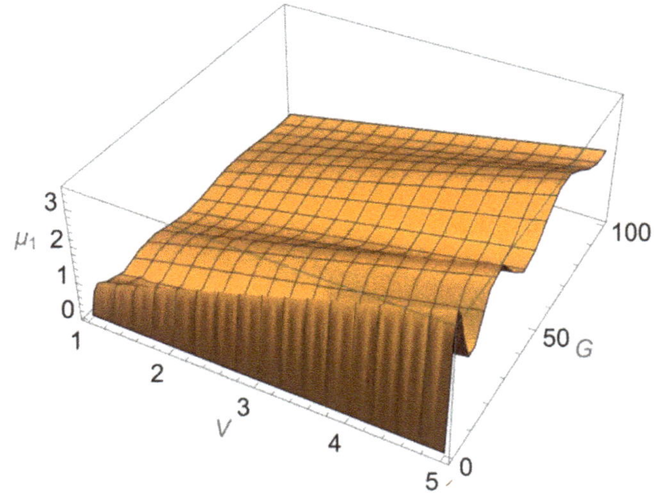

Figure 4.17. The same results as those shown in figure 4.16 but from the alternative viewpoint, so that together with figure 4.16, this figure demonstrates in a comprehensive way the nonlinear dependence of the quasienergy μ_1 on the scaled energy density of the field G and on the parameter V.

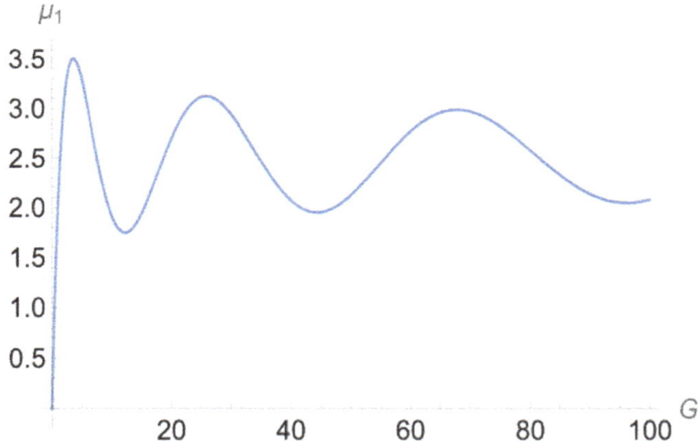

Figure 4.18. The dependence of the quasienergy μ_1 on the scaled energy density of the field G for the parameter $V = 5$.

Bibliography

[1] Baranger M and Mozer B 1961 *Phys. Rev.* **123** 25

[2] Cooper W S and Ringler H 1969 *Phys. Rev.* **179** 226

[3] Oks E 1995 *Plasma Spectroscopy: The Influence of Microwave and Laser Fields* (Berlin: Springer)

[4] Dion D R and Hirshfelder J O 1976 *Adv. Chem. Phys.* **35** 265–350

[5] Blochinzew D I 1933 *Phys. Z. Sow. Union* **4** 501

Chapter 5

Intra-Stark spectroscopy of plasmas: the nonlinear optical phenomenon and its applications

The simultaneous interaction of the radiator with the quasimonochromatic electric field and the quasistatic part of the plasma electric field led to the discovery of a new sub-area of plasma spectroscopy: *intra-Stark spectroscopy*. It deals with Langmuir-wave-induced structures (bump-dip-bump structures) at certain locations in the profiles of hydrogenic spectral lines. These structures (called for brevity Langmuir 'dips'), which consist of a local intensity minimum surrounded by two 'bumps' (peaks), constitute an *emergent phenomenon* that springs from *multifrequency nonlinear dynamic resonances* (see, e.g. papers [1, 2] and books [3, 4]). The analytical predictions of the emergent phenomenon of the Langmuir-wave-caused 'dips' were confirmed by a large number of experiments by various experimental groups working with different plasma machines as well as by astrophysical observations. In these experiments and observations, which span about ten orders of magnitude of electron densities, the highly localized Langmuir-wave-induced structures were reliably detected, identified, and used for plasma diagnostics. In particular, this included the high-precision, benchmark experiments at the gas-liner pinch by Kunze's group [5, 6], in which plasma parameters were measured using coherent Thomson scattering independently of measurements of the line profiles.

The physics behind the Langmuir-wave-induced structures is as follows. Let us consider the electric field

$$\mathbf{E}(t) = \mathbf{F} + \mathbf{E}_0 \cos(\omega t), \tag{5.1}$$

where **F** represents the quasistatic part of the electric field in the plasma. The field **F** may not only include the contribution from the quasistatic part of the ion microfield but also the contribution from the low-frequency electrostatic plasma turbulence (such as ion acoustic waves, lower hybrid waves, or Bernstein modes). If **F** and $\mathbf{E}_0$

doi:10.1088/978-0-7503-5552-0ch5

are not collinear (which is true for the overwhelming majority of possible mutual orientations of the vectors **F** and $\mathbf{E}_0$), then the total field $\mathbf{E}(t)$ is *librating*. The frequency spectrum of this librating field consists not only of the frequency ω but also its *harmonics*: the frequency spectrum of the librating field is $u\omega$, where $u = 1, 2, 3, \ldots$.

F_{eff} denotes the absolute value of the total electric field averaged over the period of the libration:

$$F_{\text{eff}} = \langle |\mathbf{E}(t)| \rangle. \tag{5.2}$$

If the librating nature of the total electric field is first disregarded, the energy levels of a radiating hydrogenic atom/ion (the radiator) of the nuclear charge Z_r split into $2n - 1$ Stark sublevels separated by (in atomic units)

$$\Omega = 3nF_{\text{eff}}/(2Z_r), \tag{5.3}$$

where n is the principal quantum number. The Stark sublevels are distinguished by the electric quantum number

$$q = n_1 - n_2, \tag{5.4}$$

where n_1 and n_2 are the parabolic quantum numbers.

The combined system 'radiator + field' can be described in terms of quasienergy states, whose quasienergies Q are as follows:

$$Q = \Omega + v\omega, \quad v = 0, \pm 1, \pm 2, \pm 3, \ldots. \tag{5.5}$$

We now take into account the time-dependent component of the librating electric field. We recall that its frequency spectrum is $u\omega$, where $u = 1, 2, 3, \ldots$. Here, we come to the central point. In the situation in which

$$\Omega = u\omega, \quad u = 1, 2, 3, \ldots, \tag{5.6}$$

there are multiple resonances between the harmonics of the librating field and *all* quasienergy states of the quasienergies $Q = \Omega + v\omega$. In other words, the resonances are multiquantum (in terms of the quanta of the Langmuir field) and multifrequency. This causes the degeneracy of all quasienergy states: the quasienergy harmonics resulting from each of the $2n - 1$ atomic Stark substates are superimposed on each other.

In this multiquantum multifrequency resonance, each degenerate quasienergy state is the superposition of several quasienergy harmonics originating from different Stark sublevels (sublevels of different values of the electric quantum number q). The Stark sublevel of some value of q is coupled by the dipole matrix element with the sublevels $q + 1$ and $q - 1$. As a result of this coupling, additional splitting of *all* quasienergy harmonics occurs. This splitting is analogous to Rabi splitting, but it is its generalization for the case of the multiquantum multifrequency resonances. The additional splitting of *all* quasienergy harmonics is generally a *nonlinear* function of the Langmuir field amplitude E_0 (see, e.g. equation (5.9) from [2]).

The multiquantum multifrequency resonances correspond to a set of specific locations in the profile of a hydrogenlike spectral line because they correspond to specific resonance values of $F_{\rm eff}$ that satisfy condition (5.5). These locations are separated from the center of the spectral line by well-defined amounts of the wavelength $\Delta\lambda^{\rm dip}(\omega)$, where $\Delta\lambda^{\rm dip}(\omega)$ are well-defined functions of the Langmuir-wave frequency $\omega = (4\pi e^2 N_e/m_e)^{1/2}$, so that (see, for example, [3, 4])

$$\Delta\lambda^{\rm dip} = aN_e^{1/2} + bN_e^{3/4}, \tag{5.7}$$

where the coefficients a and b are controlled by quantum numbers and by the charges of the radiating and perturbing ions. The identification of these structures in the experimental line profile allows the electron density N_e to be very accurately determined from their locations. From equation (5.7) it is clear that the locations of these resonance structures in the spectral line profile depend on the electron density in a nonlinear way. There is more nonlinearity in this phenomenon, as explained further in this chapter.

At each exact location in the line profile corresponding to the resonance (5.6), (for brevity, the 'resonance location'), a partial transfer of intensity takes place from the wavelength of the exact resonance location to adjacent wavelengths on each side of the exact resonance location due to the generalized Rabi splitting of the quasienergies. As a result, a structure can appear that consists of the local depression of the intensity surrounded by two relatively small 'bumps,' as illustrated in figure 5.1.

When superimposed with an inclined 'unperturbed' spectral profile, *each bump-dip-bump structure can be responsible for two local minima of the intensity*—as shown in figure 5.1—rather than just one local minimum of the intensity. The secondary minimum located at a higher intensity than the primary minimum is of no physical

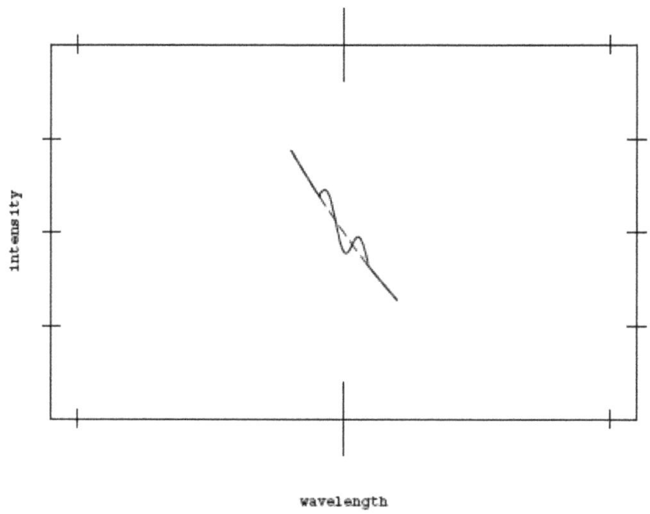

Figure 5.1. A calculated 'bump-dip-bump' structure caused by the multifrequency nonlinear dynamic resonance (5.5), superimposed on an inclined 'unperturbed' spectral line profile.

significance. Sometimes one of the two bumps and/or the secondary local minimum manifests only as a small 'shoulder.'

The analytically predicted Langmuir-wave-induced 'dips' were then revealed in a large number of various experiments around the world [5–19] and in astrophysical observations [20]. In these experiments and observations, the Langmuir-wave-induced 'dips' were reliably detected, identified, and used for plasma diagnostics in plasmas of electron densities ranging from 10^{13} cm^{-3} to 3×10^{22} cm^{-3}.

It should be noted that the nonlinear phenomenon of intra-Stark spectroscopy is analogous to the 'older' nonlinear phenomenon of intra-Doppler spectroscopy. In intra-Doppler spectroscopy, the nonlinear phenomenon manifests as localized structures within the Doppler line profile. In intra-Stark spectroscopy, the nonlinear phenomenon manifests as localized structures within the quasistatic Stark profile of the spectral line. Below are more analytical details of this phenomenon.

In order to present the analytical details in the simplest way, we focus here on the case where in the total electric field given by equation (5.1), the quasistatic field $\mathbf{F}$ is perpendicular to the quasimonochromatic field amplitude $\mathbf{E}_0$. In this case, the nonadiabatic effects are expected to be the strongest. The general case, in which the angle between $\mathbf{F}$ and $\mathbf{E}_0$ is arbitrary, is presented in paper [2] and book [3].

Let the z-axis of the fixed coordinate system lie along $\boldsymbol{E}_0$ and the x-axis lie along $\mathbf{F}$. The analysis is performed by adopting the rotating (or more rigorously, librating) reference frame whose z'-axis lies along the total electric field $\mathbf{E}(t)$ at all instants of time. The z'-axis constitutes a time-dependent angle $\varphi(t)$ with the z-axis. The Hamiltonian of the radiator in the rotating frame can be represented as follows:

$$H = H_0 + V_1(t) + V_2(t), \qquad V_1(t) = zE(t), \qquad V_2(t) = l_y \mathrm{d}\varphi/\mathrm{d}t. \tag{5.8}$$

In equation (5.8), l_y is the angular momentum projection onto the y-axis, H_0 is the Hamiltonian of the unperturbed atom in the fixed reference frame, and

$$E(t) = |\,\mathbf{E}(t)\,|, \qquad \mathrm{d}\varphi/\mathrm{d}t = \omega F E_0(\sin\omega t)/E^2(t). \tag{5.9}$$

The time-averaged value of the perturbation $V_1(t)$ splits the Stark level of the principal quantum number n into $2n - 1$ equidistant sublevels separated from each other by

$$\Delta\omega = (3n/\pi)(F^2 + A)^{1/2}E(k), \tag{5.10}$$

where

$$A = E_0{}^2 \tag{5.11}$$

is the scaled energy density of the oscillatory electric field, $E(k)$ is the complete elliptic integral of the second kind, and

$$k = A^{1/2}/(F^2 + A)^{1/2}. \tag{5.12}$$

Figure 5.2 shows the dependence of the separation $\Delta\omega$ from equation (5.10) on the scaled energy density A of the quasimonochromatic electric field and on the quasistatic field F for $n = 2$.

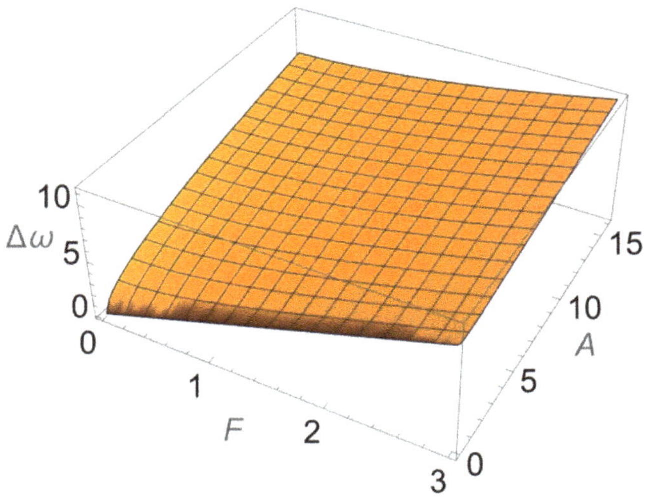

Figure 5.2. The dependence of the separation $\Delta\omega$ from equation (5.10) on the scaled energy density A of the quasimonochromatic electric field and on the quasistatic field F for $n = 2$.

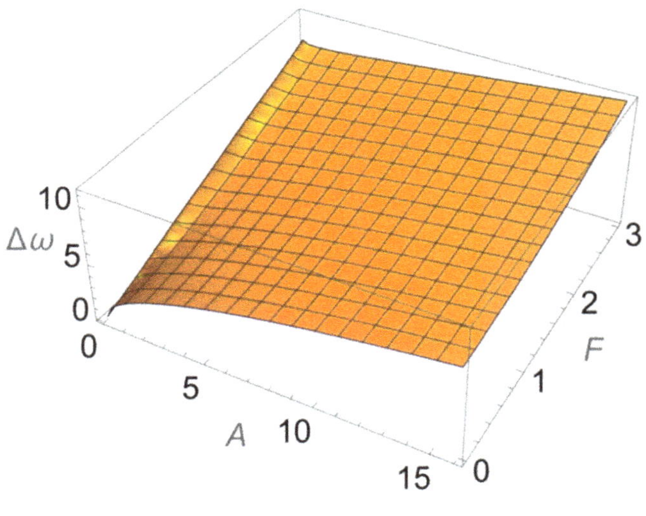

Figure 5.3. The same plot as in figure 5.2 but from the alternative viewpoint, so that together with figure 5.2 it demonstrates in the most comprehensive way the dependence of the separation $\Delta\omega$ from equation (5.10) on the scaled energy density A of the quasimonochromatic electric field and on the quasistatic field F for $n = 2$.

Figure 5.3 shows the same plot as in figure 5.2 but from the alternative viewpoint, so that together with figure 5.2 it demonstrates in the most comprehensive way the dependence of the separation $\Delta\omega$ from equation (5.10) on the scaled energy density A of the quasimonochromatic electric field and on the quasistatic field F for $n = 2$.

From figures 5.2 and 5.3, it can be seen that the separation $\Delta\omega$ depends on the scaled energy density A of the quasimonochromatic electric field in a nonlinear way.

In the particular case of one-quantum resonance, corresponding to $u = 1$ in equation (5.6), the system of equations used to determine the quasienergies λ consists of n^2 linear equations. To solve this system in a straightforward way, one should equate the determinant of this system to zero. As a result, one would have to find the roots of the polynomial of degree n^2 with respect to λ, so that for $n > 2$ there would be no analytical solution. Luckily, there is a way out, namely that this system of equations is equivalent to the system describing a static Stark effect in the effective electric field $E_0/2$. Consequently, the quasienergies can be expressed as follows (for $E_0 \ll F$):

$$\lambda = 3nq\ A^{1/2}/2, \qquad q = n_1 - n_2, \tag{5.13}$$

where A is the scaled energy density of the quasimonochromatic electric field (see equation (5.11)) and n_1 and n_2 are the parabolic quantum numbers.

Figure 5.4 presents the dependence of the quasienergies on the scaled energy density of the quasimonochromatic electric field for $n = 3$. It can be seen that the quasienergies depend on the energy density of the quasimonochromatic electric field in a nonlinear way.

Figure 5.5 presents the spectrum of the Lyman-alpha line in the mutually perpendicular fields **F** and $\boldsymbol{E}_0 \cos \omega t$ under one-quantum resonance in the $n = 2$ level. The direction of observation is perpendicular to both **F** and $\mathbf{E}_0$. The transmission axis of the linear polarizer is either parallel to **F** or parallel to **E**. The split is equal to $\Omega_0 = 3E_0/2$.

Figure 5.6 shows the same as figure 5.5 but for the Lyman-beta line for the case of one-quantum resonance in the $n = 3$ level ($\Omega_0 = 9E_0/4$).

Figure 5.7 displays the same as in figure 5.6 but for the Balmer-alpha line. The entire spectrum is presented only at $\Delta\omega \geqslant 0$, since the spectrum is symmetric with respect to the unperturbed frequency $\Delta\omega = 0$. The intensities of the components inside the *wavy lines* are shown enlarged by a factor of 200.

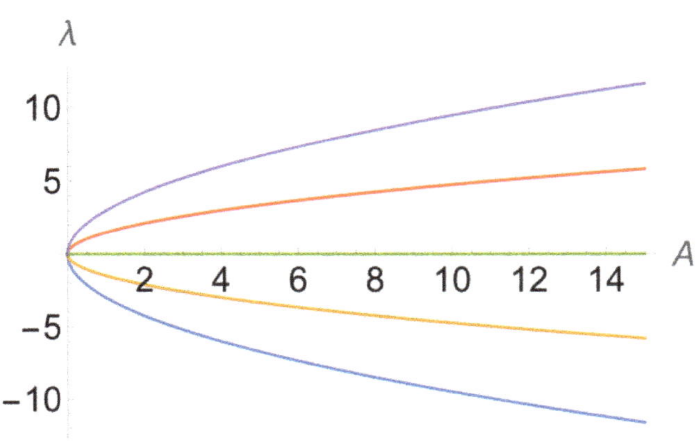

Figure 5.4. The dependence of the quasienergies on the scaled energy density of the quasimonochromatic electric field for $n = 3$ in the case of one-quantum resonance, corresponding to $u = 1$ in equation (5.6).

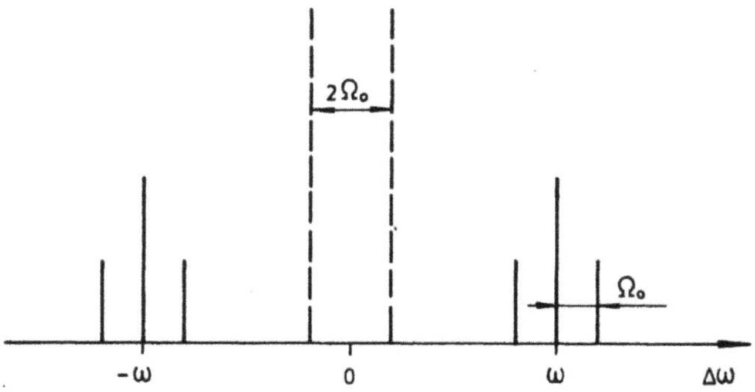

Figure 5.5. The spectrum of the Lyman-alpha line in the mutually perpendicular fields **F** and $E_0 \cos \omega t$ under one-quantum resonance in the $n = 2$ level. The direction of observation is perpendicular to both **F** and $\mathbf{E}_0$. The transmission axis of the linear polarizer is either parallel to **F** (solid lines) or parallel to **E** (dashed lines). The split is equal to $\Omega_0 = 3E_0/2$. Reprinted from [3], Copyright (1995), with permission from Springer.

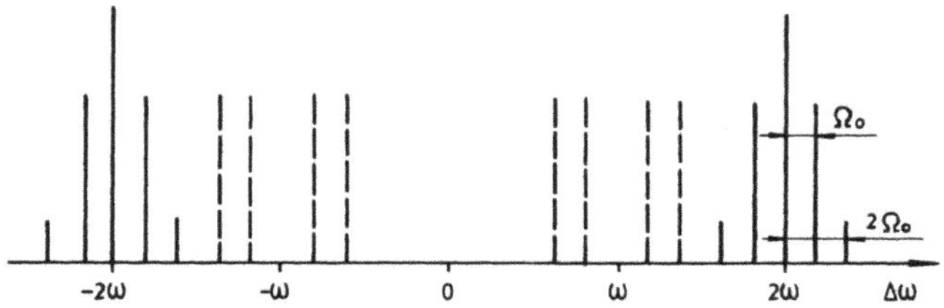

Figure 5.6. The same as figure 5.5 but for the Lyman-beta line under single-quantum resonance in the $n = 3$ level ($\Omega_0 = 9E_0/4$). Reprinted from [3], Copyright (1995), with permission from Springer.

Let us now present some details from the high-precision, benchmark experiments by Kunze's group at the gas-liner pinch [5], where plasma parameters were independently measured by both coherent Thomson scattering and measurements of the line profiles. First, this reliably established the existence of the Langmuir-wave-induced 'dips'; the evolution of their positions as the electron density varied was consistent with theory—see figure 5.8.

Figure 5.9 presents a magnified part of the profile of the Ly-alpha line obtained in the high-precision, benchmark experiment at the gas-liner pinch [5], showing the observed bump-dip-bump structure, as predicted by theory.

Figure 5.10 demonstrates that the experimental positions of the Langmuir dips yielded the values of the electron density just as accurately as the electron densities measured using coherent Thomson scattering.

During relativistic laser–plasma interactions, very strong magnetic fields of gigagauss or multi-gigagauss scales are expected to develop. Such strong magnetic fields influence the separation between Langmuir dips in the same spectral line profile. Based on this effect, [21] proposed a method for measuring these strong

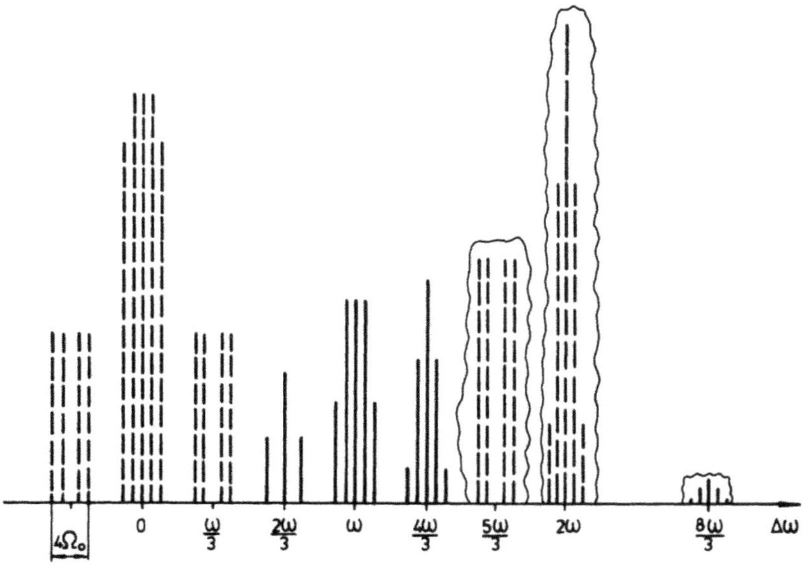

Figure 5.7. The same as in figure 5.6 but for the Balmer-alpha line. The entire spectrum is presented only at $\Delta\omega \geqslant 0$, since the spectrum is symmetric with respect to the unperturbed frequency $\Delta\omega = 0$. The intensities of the components inside the *wavy lines* are shown enlarged by a factor of 200. Reprinted from [3], Copyright (1995), with permission from Springer.

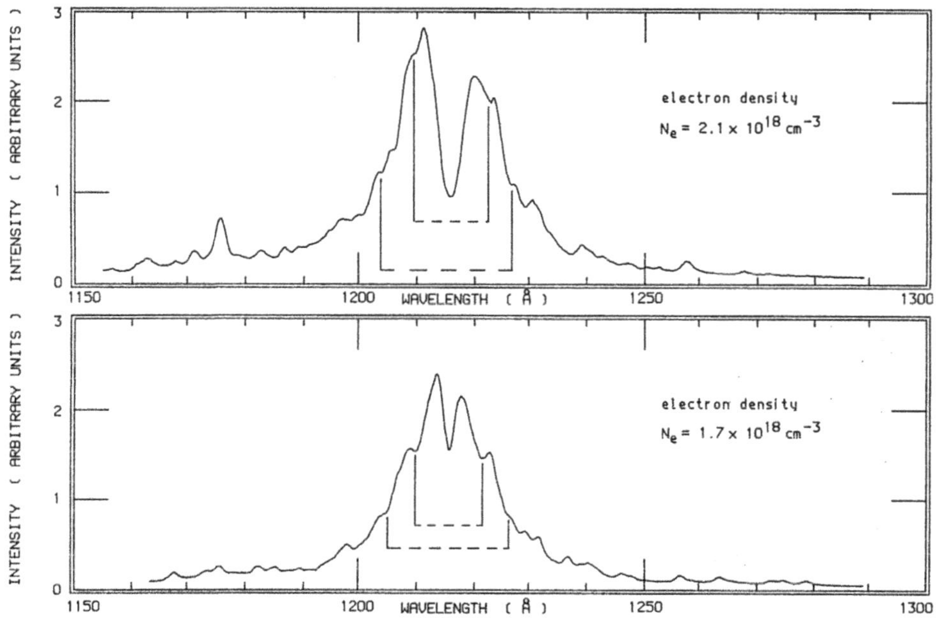

Figure 5.8. A comparison of the experimental and theoretical positions of the dips in Lyman-alpha line profiles caused by the simultaneous action of the quasistatic part of the ion microfield and the dynamic electric field of the Langmuir waves. The spectra were taken 110 ns (top) and 200 ns (bottom) after the maximum compression. The electron densities measured by coherent Thomson scattering are also indicated. The theoretical positions of the Langmuir dips are shown by pairs of vertical solid lines connected by a dashed line. Reprinted figure with permission from [5], Copyright (1991) by the American Physical Society.

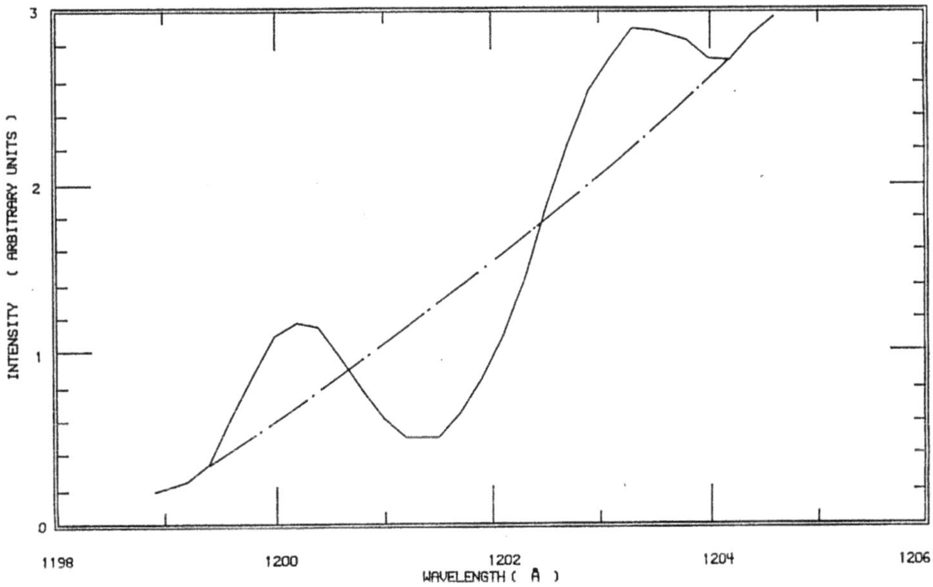

Figure 5.9. A magnified part of the profile of the Ly-alpha line obtained in the high-precision, benchmark experiment at the gas-liner pinch, showing the observed bump-dip-bump structure, as predicted by theory. Reprinted figure with permission from [5], Copyright (1991) by the American Physical Society.

magnetic fields using non-Lyman lines, e.g. Balmer lines in the x-ray range. The authors of [22] proposed an alternative method based on the effect of the strong magnetic fields on the width of the Langmuir dips.

Figure 5.11 shows the theoretical ratio $\Delta\omega_{\text{dip}}^{(\beta)}/\Delta\omega_{\text{dip}}^{(\alpha)}$ of the dip positions in a pair of Langmuir-wave-induced dips for the Balmer-alpha line and the Balmer-beta line versus the scaled dimensionless magnetic field b defined as follows:

$$b = \mu_0 B/(s\hbar\omega) = (1/s)[B(GG)/0.201][\omega(\text{sec}^{-1})/(1.77 \times 10^{15})]^{-1}, \tag{5.14}$$

where μ_0 is the Bohr magneton and s is the number of the Langmuir-wave quanta involved in the resonance that led to the dips.

The authors of [22] introduced the following three dimensionless quantities:

$$f = \hbar F/(Z_r m_e e\omega), \qquad \gamma = \hbar E_0/(Z_r m_e e\omega), \qquad b_0 = \mu_0 B/(\hbar\omega), \tag{5.15}$$

where Z_r is the nuclear charge of the radiating hydrogenlike ion. Figure 5.12 displays the theoretical dependence of the scaled width w of the Langmuir dips in the case of one-quantum resonance on the scaled magnetic field b_0 at $\gamma = 0.01$ for the following three values of the scaled quasistatic electric field: $f = 0.03$, $f = 0.1$, and $f = 0.3$.

Figure 5.13 demonstrates the same results as figure 5.12 but for the case of two-quantum resonance.

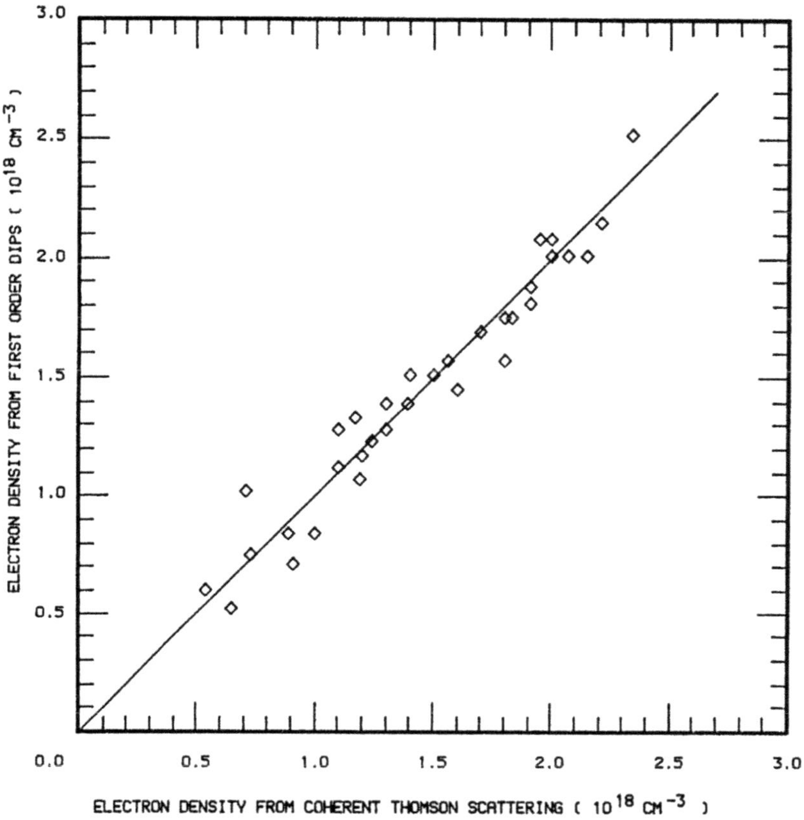

Figure 5.10. Electron densities obtained from the experimental positions of the Langmuir-wave-induced bump-dip-bump structures versus the electron densities measured using coherent Thomson scattering. The comparison is shown for many electron densities in experiments at the gas-liner pinch. Reprinted figure with permission from [5], Copyright (1991) by the American Physical Society.

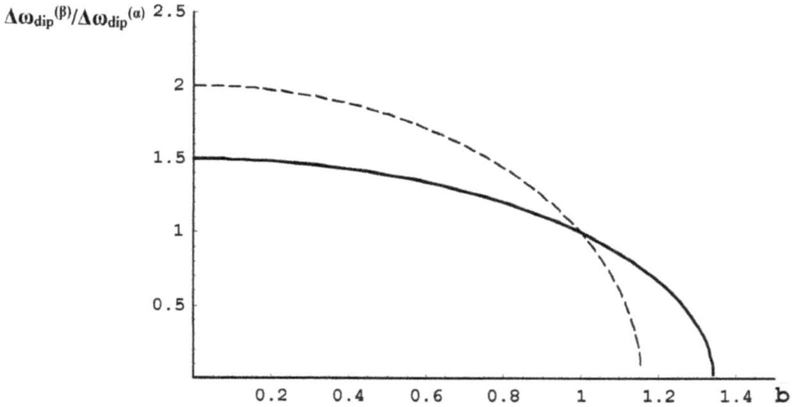

Figure 5.11. The theoretical ratio $\Delta\omega_{dip}^{(\beta)}/\Delta\omega_{dip}^{(\alpha)}$ of the dip positions in a pair of Langmuir-wave-induced dips for the Balmer-alpha line (solid curve) and for the Balmer-beta line (dashed curve) versus the scaled dimensionless magnetic field b defined in the above equation (5.14). Reprinted from [22], Copyright (2019), with permission from Elsevier.

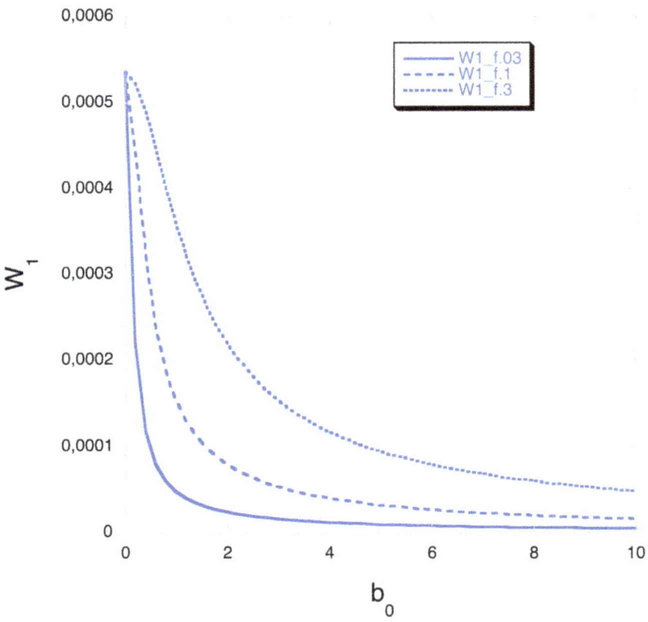

Figure 5.12. The theoretical dependence of the scaled width w of the Langmuir dips in the case of one-quantum resonance on the scaled magnetic field b_0 at $\gamma = 0.01$ for the following three values of the scaled quasistatic electric field: $f = 0.03$ (solid line), $f = 0.1$ (dashed line), and $f = 0.3$ (dotted line). The dimensionless quantities b_0, γ, and f are defined in the above equation (5.15). Reprinted from [22], Copyright (2019), with permission from Elsevier.

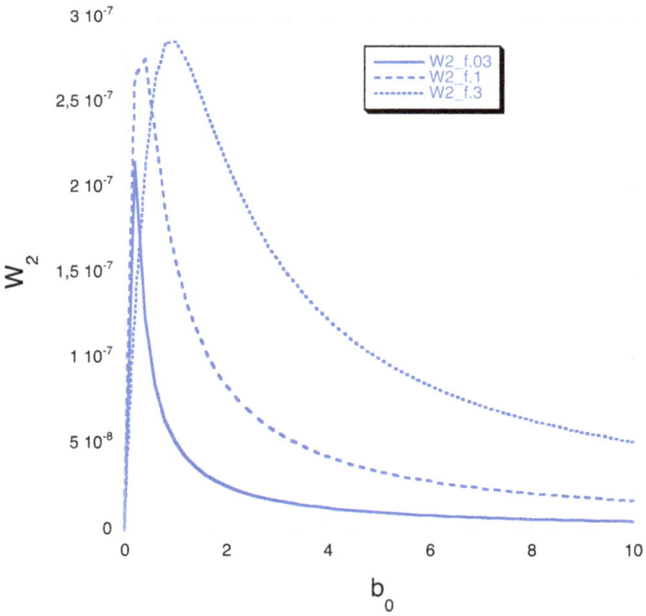

Figure 5.13. The same results as those of figure 5.12 but for the case of two-quantum resonance. Reprinted from [22], Copyright (2019), with permission from Elsevier.

Bibliography

[1] Gavrilenko V P and Oks E 1981 *Sov. Phys. JETP* **53** 1122 http://jetp.ras.ru/cgi-bin/e/index/e/53/6/p1122?a=list

[2] Gavrilenko V P and Oks E 1987 *Sov. J. Plasma Phys.* **13** 22

[3] Oks E 1995 *Plasma Spectroscopy: The Influence of Microwave and Laser Fields* (Berlin: Springer)

[4] Oks E 2017 *Diagnostics of Laboratory and Astrophysical Plasmas Using Spectral Lineshapes of One-, Two-, and Three-Electron Systems* (Singapore: World Scientific)

[5] Oks E, Böddeker St and Kunze H-J 1991 *Phys. Rev.* A **44** 8338

[6] Bertschinger G 1980 Messungen von VUV Linien an einem dichten Z-Pinch-Plasma *PhD Thesis* Bochum Ruhr-University

[7] Gavrilenko V P, Faenov A Y, Magunov A I, Pikuz T A, Skobelev I Y, Kim K Y and Milchberg H M 2006 *Phys. Rev.* A **73** 013203

[8] Renner O, Dalimier E, Oks E, Krasniqi F, Dufour E, Schott R and Förster E 2006 *J. Quant. Spectrosc. Radiat. Transfer* **99** 439

[9] Woolsey N, Howe J, Chambers D, Courtois C, Förster E, Gregory C, Hall I, Renner O and Uschmann I 2007 *High Energy Density Phys.* **3** 292

[10] Oks E, Dalimier E and Faenov A Y *et al* 2017 *Opt. Express* **25** 1958

[11] Oks E, Dalimier E and Faenov A Y *et al* 2017 *J. Phys. B: At. Mol. Opt. Phys.* **50** 245006

[12] Zhuzhunashvili A I and Oks E 1977 *Sov. Phys. JETP* **46** 1122 http://jetp.ras.ru/cgi-bin/e/index/e/46/6/p1122?a=list

[13] Dalimier E, Oks E and Renner O 2014 *Atoms* **2** 178

[14] Dalimier E, Pikuz T A and Angelo P 2018 *Atoms* **6** 45

[15] Oks E and Rantsev-Kartinov V A 1980 *Sov. Phys. JETP* **52** 50 http://jetp.ras.ru/cgi-bin/e/index/e/52/1/p50?a=list

[16] Jian L, Shali X, Quingguo Y, Lifeng L and Yufen W 2013 *J. Quant. Spectrosc. Radiat. Transfer* **116** 41

[17] Oks E, Dalimier E and Faenov A Y *et al* 2014 *J. Phys. B: At. Mol. Opt. Phys.* **47** 221001

[18] Oks E, Dalimier E and Faenov A Y *et al* 2017 *J. Phys.: Conf. Ser.* **810** 012004

[19] Dalimier E, Oks E and Renner O 2017 *AIP Conf. Proc.* **1811** 190003

[20] Oks E 1978 *Sov. Astron. Lett.* **4** 223

[21] Dalimier E and Oks E 2018 *Atoms* **6** 60

[22] Oks E, Dalimier E and Angelo P 2019 *Spectrochim. Acta* B **157** 1

Chapter 6

Nonlinear effects in the spectrum of the quasistatically Stark-broadened spectral lines of plasma under a high-frequency laser field

The effect of the *dynamic* electric field

$$\boldsymbol{E}(t) = \boldsymbol{E}_0 \cos \omega t + \boldsymbol{F} \tag{6.1}$$

(where **F** represents the quasistatic field in a plasma) on a hydrogenic atom/ion of the nuclear charge Z can be reduced to the Stark effect in the following effective *static* field (if the field $\mathbf{E}_0 \cos \omega t$ is either strong or has a high frequency):

$$\boldsymbol{F}_{\text{eff}}^{(n)} = (F_x J_0(3nE_0/2Z\omega),\ 0,\ F_z) \tag{6.2}$$

in the basis of the parabolic wave functions with the quantization axis Oz along the vector $\mathbf{E}_0$ [1]. In equation (6.2), n is the principal quantum number. Physically, this means that the high-frequency laser field $\mathbf{E}_0 \cos \omega t$ suppresses—partially or completely—the components of the quasistatic field perpendicular to $\mathbf{E}_0$. (This discussion uses atomic units.)

The corresponding radiation spectrum of the Ly-alpha line in the x-polarization is as follows [1]:

$$\begin{aligned} I(\Delta\omega, E_0) = {} & [F_z^2/(2F_{\text{eff}}^2)]\delta(\Delta\omega) \\ & + \{F_x^2 J_0^2[3nE_0/(2Z\omega)]/F_{\text{eff}}^2\} \\ & [\delta(\Delta\omega - 3F_{\text{eff}}/Z) + \delta(\Delta\omega + 3F_{\text{eff}}/Z)], \end{aligned} \tag{6.3}$$

where $\delta(\cdots)$ are the Dirac delta functions. So, the spectrum consists of the unshifted component of the intensity $[F_z^2/(2F_{\text{eff}}^2)]$ and the two components of the intensity $F_x^2 J_0^2[3nE_0/(2Z\omega)]/F_{\text{eff}}^2$ shifted by $\pm 3F_{\text{eff}}/Z$.

We introduce the following scaled dimensionless energy density of the laser field

$$A = [3nE_0/(2Z\omega)]^2 \tag{6.4}$$

and the ratio of the F_x and F_z components of the quasistatic field:

$$R = F_x/F_z. \tag{6.5}$$

Figure 6.1 shows the dependence of the intensity I_u of the unshifted component on the scaled energy density of the field A and on the ratio R.

Figure 6.2 shows the same results as those shown in figure 6.1 but from the alternative viewpoint, so that together with figure 6.1, it demonstrates in a comprehensive way the

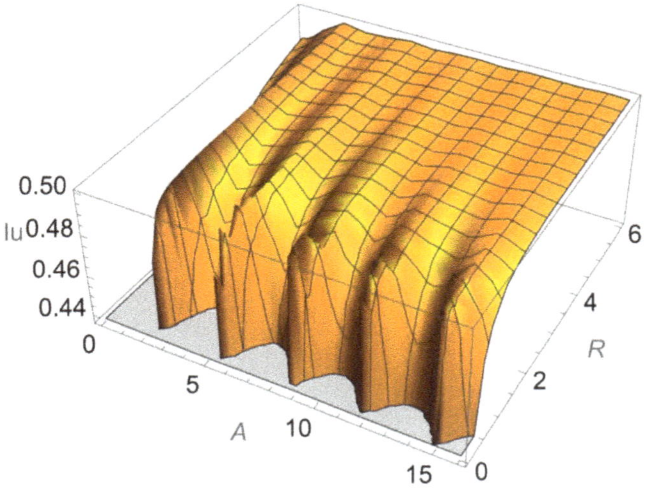

Figure 6.1. The dependence of the intensity I_u of the unshifted component on the scaled energy density of the field A and on the ratio R.

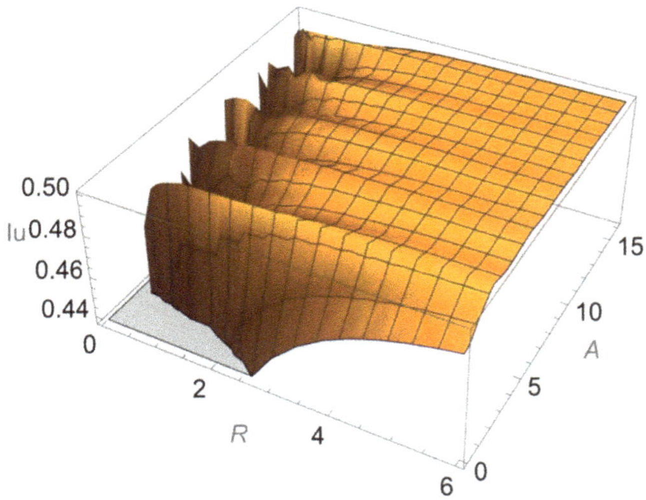

Figure 6.2. The same results as those shown in figure 6.1 but from the alternative viewpoint, so that together with figure 6.1, this figure demonstrates in a comprehensive way the nonlinear dependence of the intensity of the unshifted component on the scaled energy density of the field at various values of the ratio $R = F_x/F_z$.

nonlinear dependence of the intensity of the unshifted component on the scaled energy density of the field at various values of the ratio $R = F_x/F_z$.

Figure 6.3 shows the dependence of the intensity I_u of the unshifted component on the scaled energy density of the field A at the ratio $R = 1$.

Figure 6.4 shows the dependence of the intensity I_s of any of the two shifted components on the scaled energy density of the field A and on the ratio R.

Figure 6.5 shows the same results as those shown in figure 6.4 but from the alternative viewpoint, so that together with figure 6.4, it demonstrates in a comprehensive way the nonlinear dependence of the intensity of any of the two shifted components on the scaled energy density of the field at various values of the ratio $R = F_x/F_z$.

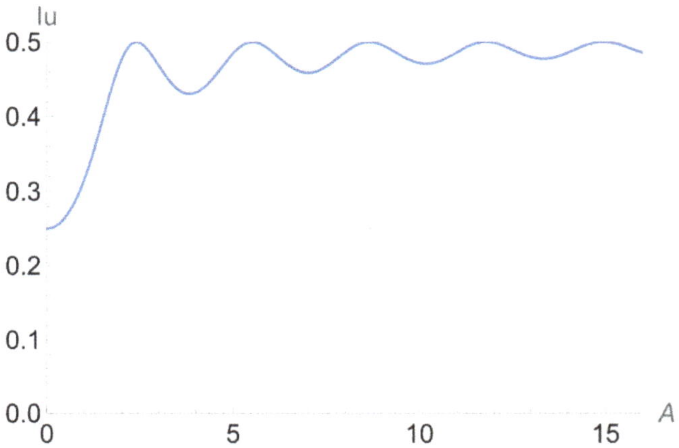

Figure 6.3. The dependence of the intensity I_u of the unshifted component on the scaled energy density of the field A at the ratio $R = 1$.

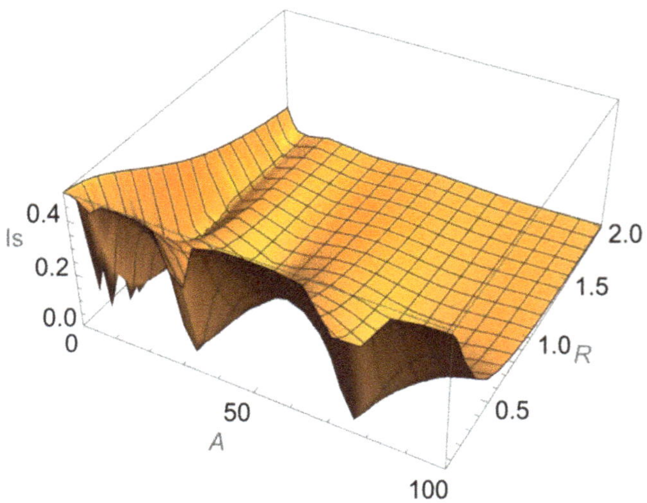

Figure 6.4. The dependence of the intensity I_s of any of the two shifted components on the scaled energy density of the field A and on the ratio R.

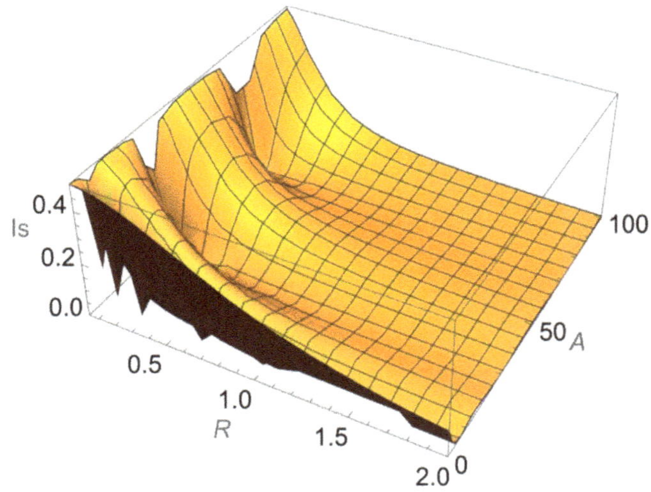

Figure 6.5. The same results as those shown in figure 6.4 but from the alternative viewpoint, so that together with figure 6.4, this figure demonstrates in a comprehensive way the nonlinear dependence of the intensity of any of the two shifted components on the scaled energy density of the field at various values of the ratio $R = F_x/F_z$.

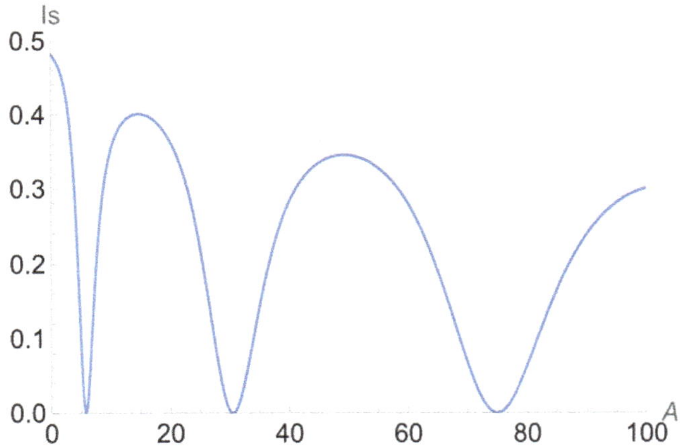

Figure 6.6. The dependence of the intensity I_s of any of the two shifted components on the scaled energy density of the field A at the ratio $R = 0.2$.

Figure 6.6 shows the dependence of the intensity I_s of any of the two shifted components on the scaled energy density of the field A at the ratio $R = 0.2$.

Figure 6.7 shows the dependence of the shift S of the blue-shifted component on the scaled energy density of the field A and on the ratio R.

Figure 6.8 shows the same results as those shown in figure 6.7 but from the alternative viewpoint, so that together with figure 6.7, it demonstrates in a comprehensive way the nonlinear dependence of the shift of the blue-shifted component on the scaled energy density of the field at various values of the ratio $R = F_x/F_z$.

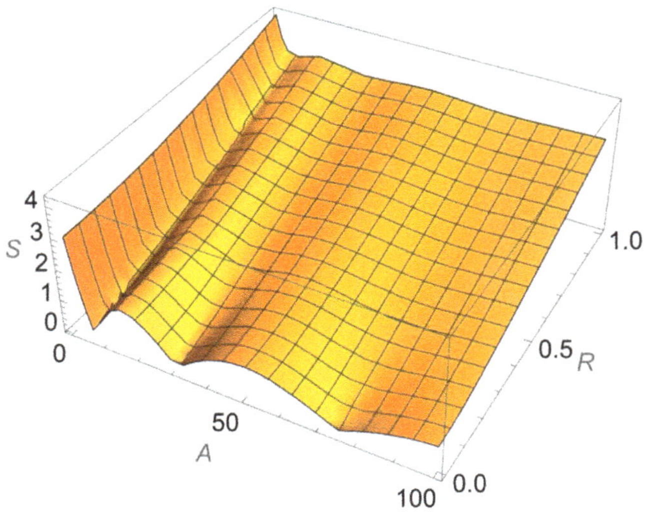

Figure 6.7. The dependence of the shift S of the blue-shifted component on the scaled energy density of the field A and on the ratio R.

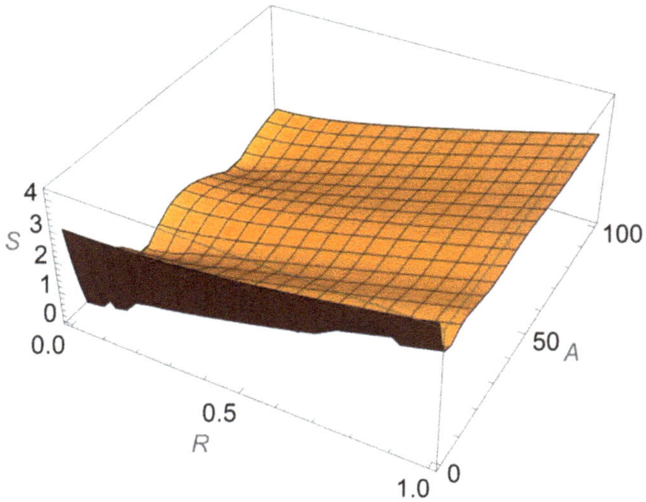

Figure 6.8. The same results as those shown in figure 6.7 but from the alternative viewpoint, so that together with figure 6.7, this figure demonstrates in a comprehensive way the nonlinear dependence of the shift of the blue-shifted component on the scaled energy density of the field at various values of the ratio $R = F_x/F_z$.

Figure 6.9 shows the dependence of the shift S of the blue-shifted component on the scaled energy density of the field A at the ratio $R = 0.2$.

From figures 6.1–6.9, it can be seen that the intensities of the shifted and unshifted components of the Ly-alpha line and the shift of the blue-shifted component depend on the scaled energy density of the high-frequency laser field in a highly nonlinear way.

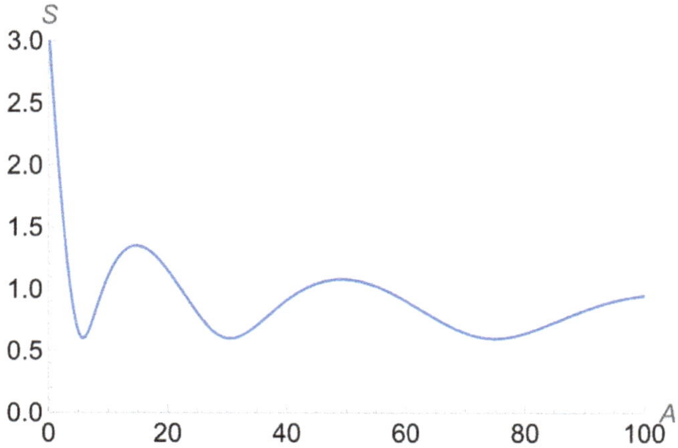

Figure 6.9. The dependence of the shift S of the blue-shifted component on the scaled energy density of the field A at the ratio $R = 0.2$.

Bibliography

[1] Oks E 1995 *Plasma Spectroscopy: The Influence of Microwave and Laser Fields* (Berlin: Springer)

Eugene Oks

Chapter 7

Nonlinear effects in the radiation of dynamically-Stark-broadened hydrogenic spectral lines by plasma electrons

7.1 Early theoretical results

In the early analytical theories of the Stark broadening of spectral lines by the electron microfield in plasmas, the electrons were considered in the dynamic regime. The early theories disregarded the back reaction of the radiating atom/ion (hereafter, the radiator) on the perturbing electrons, i.e. they were developed in the approximation of 'no back reaction.' In the frame of this approximation, there are two types of analytical theory: semiclassical theories and limited-quantal theories. In the semiclassical theories, the perturbing electrons are described as classical particles, whereas the radiator is described quantum mechanically. In the limited-quantal theories, both the perturbing electrons and the radiator are described quantum mechanically.

The first semiclassical theory was the so-called 'conventional theory' (sometimes called the 'standard theory'), in which the dynamic Stark broadening due to the perturbing electrons was considered in the impact approximation [1, 2]. In a simplified way, it can be stated that in the impact approximation, the influence of the perturbing electrons was treated as a sequence of binary collisions, in which each collision was assumed to be completed. An approximate solution of the Schrödinger equation for the radiating atomic electron was then found in the second order of Dirac's perturbation theory. Some additional details of the impact approximation are presented in appendix C.

The so-called 'unified theory' [3, 4] was developed later. Its principal distinction (compared to the conventional theory) was that it took into account incomplete collisions.

For completeness, we also mention the so-called relaxation theory [5, 6]. Its central point is some integro-differential equation (rather than the Schrödinger

doi:10.1088/978-0-7503-5552-0ch7

equation) that describes how the radiating atomic electron evolves over time on average. The mathematical tools of the relaxation theory are Green's function and Zwanzig's projection operators. However, in order to obtain an analytical solution, it is necessary to make almost the same simplifying suppositions as those of the conventional theory, with the exception of assuming that the collisions are complete. Therefore, the practical version of the relaxation theory is basically the same as the unified theory.

An analytical solution of the Stark broadening problem—the exact solution beyond the perturbation theory—was later found [7] in the binary approximation. Further, in [8], the binary assumption was eliminated and an exact analytical solution was obtained in the multiparticle (i.e. the most general) picture of the interaction of the radiator with the electron microfield.

7.2 The refinement of the conventional theory for hydrogen lines

In [9], we refined the conventional theory of the Stark broadening of hydrogen lines by plasma electrons by allowing for the fact that the overwhelming majority of the hydrogen atom states have nonzero electric dipole moments:

$$e\langle \mathbf{R} \rangle = -3e^3\mathbf{A}/(4|E_{\mathrm{at}}|). \tag{7.1}$$

In equation (7.1), E_{at} and $\langle \mathbf{R} \rangle$ are the energy and the average value of the radius vector of the radiating atomic electron, and **A** is the conserved Runge–Lenz vector:

$$\mathbf{A} = -\mathbf{R}/R + (\mathbf{p}x\mathbf{L} - \mathbf{L}x\mathbf{p})/(2m_e e^2). \tag{7.2}$$

In equation (7.2), m_e, e, **L**, and **p** are the mass, charge, angular momentum, and linear momentum of the radiating atomic electron, respectively.

Since the Runge–Lenz vector **A** and the dipole moment are antiparallel, it follows that the motion of the perturbing electrons occurs in the following dipole potential:

$$V = e^2\langle \mathbf{R} \rangle \bullet \mathbf{r}/r^3. \tag{7.3}$$

In equation (7.3), **r** is the radius vector of the perturbing electrons. Thus, in distinction to the conventional theory, in which the perturbing electrons were considered to move as free particles, in the refined conventional theory, they move in the potential given by equation (7.3).

Another refinement made in [9] concerned the so-called Weisskopf radius (defined below). Griem, in his description of the conventional theory presented in his book [10], proposed a modification of the so-called strong collision constant without making the corresponding change in the Weisskopf radius. In [9], it was demonstrated that the choices of the strong collision constant and of the Weisskopf radius are interconnected.

These refinements of the conventional theory are important for plasmas that have the following parameters (called warm dense plasmas): a temperature T of a few eV (or less) and the electron density $N_e \sim 10^{18}$ cm^{-3} (or greater). Below are some details from paper [9].

Equation (7.3) can be represented in the form

$$V(r,\,\theta) = (d\cos\theta)/r^2,\quad d = e^2\,|\langle \mathbf{R}\rangle|. \tag{7.4}$$

Here, θ is the polar angle of the vector $\mathbf{r}$; the polar axis is parallel to the vector $\langle \mathbf{R}\rangle$, whose absolute value is [11]: $|\langle \mathbf{R}\rangle| = 3nqa_B/2$, where n is the principal quantum number; $q = n_1 - n_2$ is the difference between the parabolic quantum numbers; and a_B is the Bohr radius. Thus, $d = 3n\,|\,q\,|\,e^2a_B/2$ for the spectral lines of the Lyman series. For the non-Lyman lines of hydrogen, it is appropriate to use the corresponding arithmetic average for the lower $(n',\,q')$ and upper $(n,\,q)$ Stark sublevels involved in the radiative transition:

$$d = 3(n\,|\,q\,| + n'|q'|)e^2a_B/4. \tag{7.5}$$

The motion in the potential of the type given by equation (7.4) has been analyzed by several authors [12–14]. An important property of this type of potential is the presence of an additional conserved quantity B (in addition to the conservation of the projection M_z of the perturbing electron angular momentum $\mathbf{M}$):

$$B = M^2 + 2m_e d\cos\theta. \tag{7.6}$$

This quantity is conserved at any instant of time. Consequently, it can be represented via its asymptotic value at $t = -\infty$ as follows:

$$B = (m_e v_0 \rho_0)^2 + 2m_e d\cos\theta_0. \tag{7.7}$$

In equation (7.7), θ_0 is the angle between $\mathbf{r}(-\infty)$ and $\langle \mathbf{R}\rangle$; ρ_0 is the asymptotic impact parameter at $t = -\infty$. The angle θ_0 is the same as the angle between $-\mathbf{v}_0$ and $\langle \mathbf{R}\rangle$, $\mathbf{v}_0$ being the perturbing electron asymptotic velocity $\mathbf{v}(-\infty)$ at $t = -\infty$.

Another essential property of the physical system under consideration is the possibility of separating the angular and radial motions, so that the radial motion occurs in the effective potential

$$U_{\mathrm{eff}} = B/(2m_e r^2). \tag{7.8}$$

As a result, one obtains the following formula for the dependence of the absolute value of the perturbing electron radial coordinate on time:

$$r(t) = [B/(2m_e E) + (2E/m_e)t^2]^{1/2}. \tag{7.9}$$

In equation (7.9), E is the energy of the perturbing electron. On substituting the expression for B from equation (7.7) and the expression $E = m_e\, v_0^2/2$ in equation (7.9), we find that

$$r(t) = \left(\rho_{\mathrm{eff}}{}^2 + v_0{}^2t^2\right)^{1/2} \tag{7.10}$$

with

$$\rho_{\mathrm{eff}} = \left[\rho_0{}^2 + 2d\cos\theta_0/(m_e v_0{}^2)\right]^{1/2}. \tag{7.11}$$

Below, the subscript 0 of the quantity v_0 is omitted for brevity.

We introduced the notation

$$2d/(m_e v^2) = \rho_d^{\,2} \tag{7.12}$$

and represented equation (7.11) as follows:

$$\rho_{\text{eff}} = \left(\rho_0^{\,2} + \rho_d^{\,2}\cos\theta_0\right)^{1/2}. \tag{7.13}$$

We then applied the method of effective trajectories employed in atomic physics in calculations of cross-sections [15]. In the frames of this method, the nonrectilinear (actual) trajectory of the perturbing electrons was replaced by a rectilinear trajectory corresponding to the effective impact parameter ρ_{eff} from equation (7.13) and the velocity v_0.

The Stark width caused by plasma electrons is controlled by the diagonal elements $\langle\alpha\,|\langle\beta\mid\sigma(v)\mid\beta\rangle|\,\alpha\rangle$ of the operator $\sigma(v)$, whose physical meaning is the cross-section of the 'optical collisions' (i.e. the collisions responsible for the Stark broadening). Here, α and β are the Stark sublevels of the upper and lower energy levels (involved in the radiative transition), respectively.

After some lengthy calculations (presented in full in our paper [9]), we obtained the following expression for the ratio of the Stark width in the refined conventional theory to the corresponding Stark width of the conventional theory:

$$\begin{aligned}&\text{WIDTH RATIO}\\ &=\{\ln[(\exp(2b^2)-1)^{1/2}(1/x^4-1)^{1/4}/2^{1/2}]-b^2/2\\ &\quad+[1/(4x^2)]\ln[(1+x^2)/(1-x^2)]\}/\{\ln[b/(xC^{1/2})]+0.356\},\end{aligned} \tag{7.14}$$

where contains two dimensionless parameters:

$$x = \langle\rho_d\rangle_{av}/\rho_{max}, \tag{7.15}$$

$$b = (2C/3)^{1/2}(n^2 + n'^2)^{1/2}/(n^2 - n'^2). \tag{7.16}$$

In equation (7.15), C is the strong collision constant, whose value should be chosen to be between zero and two, and

$$\langle\rho_d\rangle_{av} = [(n^2 + n'^2)/2]^{1/2}\hbar/(m_e v), \tag{7.17}$$

while ρ_{max} is typically the Debye radius

$$\rho_D = [T_e/(4\pi e^2 N_e)]^{1/2}. \tag{7.18}$$

We note that the parameter b defined in equation (7.16) is essentially just a combination of the quantum numbers characterizing (and thus, identifying) the particular hydrogen line.

Figure 7.1 shows the dependence of the WIDTH RATIO from equation (7.14) on the parameters x and b.

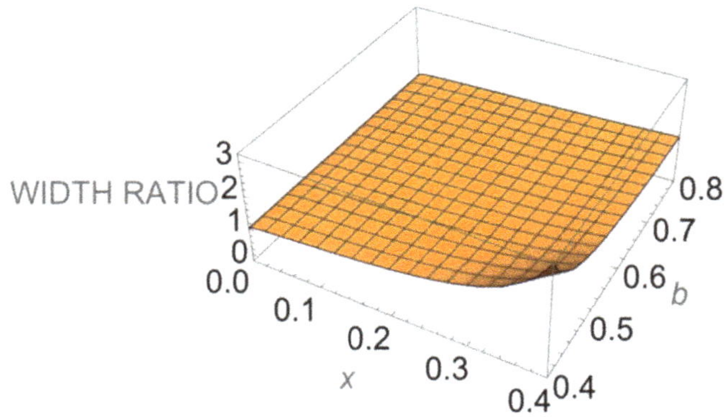

Figure 7.1. The dependence of the WIDTH RATIO from equation (7.15) on the parameters x and b defined in equation (7.16) at a strong collision constant value of $C = 3/2$. Reprinted from [9], Copyright (2015), with permission from Elsevier.

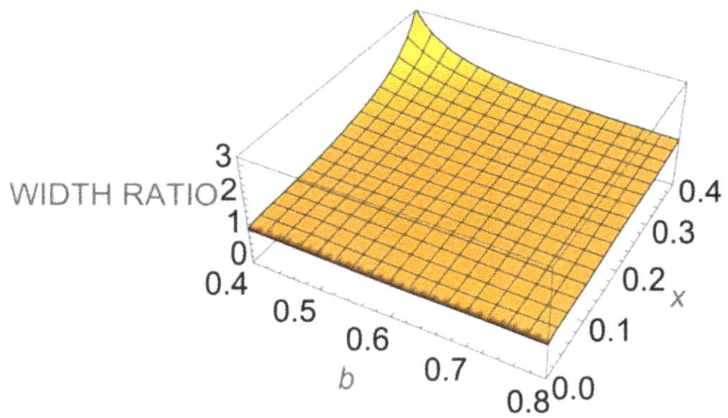

Figure 7.2. The same plot as in figure 7.1, but from the alternative viewpoint, so that together with figure 7.1 it shows the width ratio versus the parameters x and b in the most comprehensive way.

Figure 7.2 represents the same plot as in figure 7.1, but from the alternative viewpoint, so that together with figure 7.1 it shows the width ratio versus the parameters x and b in the most comprehensive way.

Figure 7.3 presents the ratio of the Stark widths from equation (7.15) for the Balmer-alpha hydrogen line versus the dimensionless parameter x (defined in equation (7.16)) for the following values of the strong collision constant: $C = 2$, $C = 3/2$, and $C = 1$ (figure 11.2 from my seventh book 'Analytical Advances in Quantum and Celestial Mechanics: Separating Rapid and Slow Subsystems').

From figures 7.1–7.3, one can see that in the refined conventional theory, the Stark width due to plasma electrons is greater than the corresponding Stark width of the conventional theory.

We now focus on the nonlinear dependence of the electron-induced Stark width on the electron density N_e in the refined conventional theory. Since the conventional

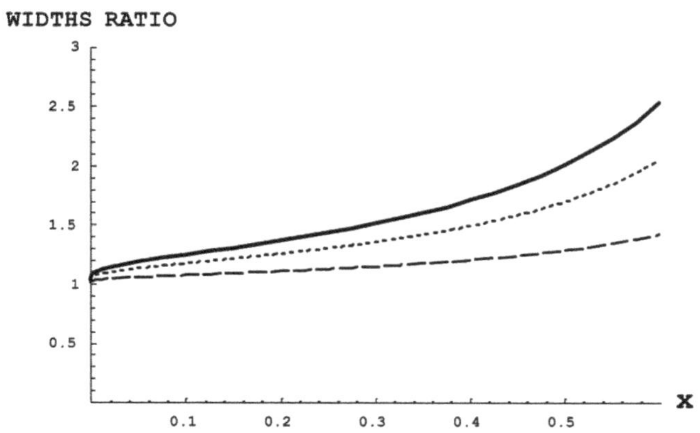

Figure 7.3. The ratio of the Stark widths from equation (7.15) for the Balmer-alpha line of hydrogen versus the dimensionless parameter *x* (defined in equation (7.16)) for the following values of the strong collision constant: $C = 2$ (solid curve), $C = 3/2$ (dotted curve), and $C = 1$ (dashed curve). Reproduced from [16].

theory treats this broadening using a sequence of *binary* encounters with the perturbing electrons, the corresponding Stark width might be expected to be *linear* with respect to the electron density. However, in reality, it is *nonlinear*, as presented below.

A more explicit expression for the parameter x from equation (7.16) is

$$x = (2e\hbar/T)[(n^2 + n'^2)N_e/m_e]^{1/2}. \tag{7.19}$$

We introduce the scaled electron density $N_{e,s}$ as follows:

$$N_{e,s} = x^2 = (2e\hbar/T)^2(n^2 + n'^2)N_e/m_e. \tag{7.20}$$

The scaled electron-induced Stark width in the refined conventional theory W_s can now be represented in the form

$$W_s = N_{e,s}\{\ln[(\exp(2b^2) - 1)^{1/2}(1/N_{e,s}^2 - 1)^{1/4}/2^{1/2}] - b^2/2 + [1/(4N_{e,s})]\ln[(1 + N_{e,s})/(1 - N_{e,s})]\}. \tag{7.21}$$

Figure 7.4 shows the dependence of the scaled electron-induced Stark width in the refined conventional theory W_s on the scaled electron density $N_{e,s}$ and on the parameter b.

Figure 7.5 represents the same plot as in figure 7.4, but from the alternative viewpoint, so that together with figure 7.4 it shows the scaled Stark width versus the scaled electron density $N_{e,s}$ and the parameter b in the most comprehensive way.

From figures 7.4 and 7.5, it can be seen that in the refined conventional theory, the electron-induced Stark width depends on the electron density in a nonlinear way, even though the interaction of the radiator with the perturbing electrons was considered using a sequence of binary encounters.

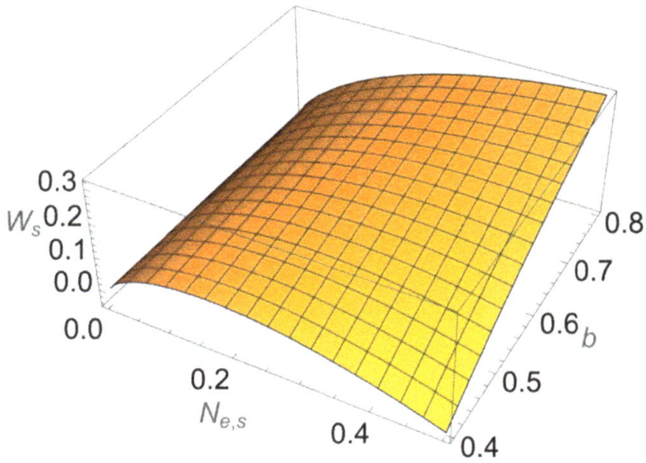

Figure 7.4. The dependence of the scaled electron-induced Stark width in the refined conventional theory W_s on the scaled electron density $N_{e,s}$ and on the parameter b.

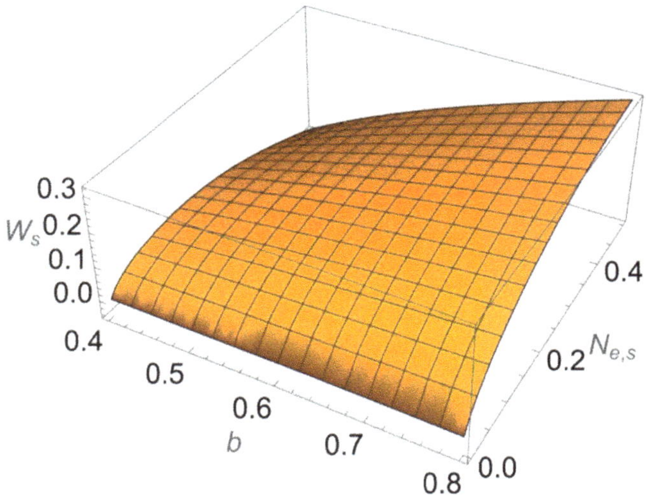

Figure 7.5. The same plot as in figure 7.4, but from the alternative viewpoint, so that together with figure 7.4 it shows the scaled Stark width versus the scaled electron density $N_{e,s}$ and the parameter b in the most comprehensive way.

7.3 The refinement of the conventional theory for hydrogenlike lines

The so-called conventional theory of the electron-induced Stark broadening of hydrogenlike spectral lines in plasmas was worked out in [17] and later also presented in [10]. Later advances can be found in [18, 19] and in references therein. The conventional theory uses a simplified picture of the corresponding interaction: there is a coulomb field that has an effective charge of $Z - 1$, and the perturbing electrons follow hyperbolic paths in this coulomb field. (Here, Z is the radiator nuclear charge; atomic units are used.) Thus, the conventional theory is based on a simplification, i.e. a two-body problem (the charge $Z - 1$ plus the perturbing electron).

However, the actual system represents a three-body problem: the nucleus, the bound electron, and the perturbing electrons. Therefore, the actual path of the perturbing electron is more complex.

In [20], we treated this three-body problem by applying the standard analytical method based on the separation of rapid and slow subsystems—see, e.g. [21]. The application of this analytical method was possible because the characteristic frequency of the 'rotation' of the perturbing electron around the radiator is much smaller than the characteristic frequency of the motion of the bound electron. Therefore, the latter can be treated as the rapid subsystem and the former can be treated as the slow subsystem. We note that this analytical method yields sufficiently accurate results in cases in which the standard perturbation theory is invalid—see, e.g. [21].

Thus, in paper [20] we produced significantly more accurate analytical results (for the operator of the Stark broadening by electrons) than those produced by the conventional theory.

We demonstrated that, for instance, for Lyman spectral lines of ionized helium (He II), the three-body treatment increases the electron-induced Stark width, and that the higher the electron density N_e, the greater this increase becomes. Some details are given below.

The conventional theory starts from the following expression for the operator of the Stark broadening by electrons [16]:

$$\Phi_{ab} \equiv 2\pi v\, N_e \int d\rho\, \rho\, \{S_a S_b^* - 1\}, \tag{7.22}$$

where ρ and v are the impact parameter and the velocity of the perturbing electron, respectively. In equation (7.22), $S_a(0)$ and $S_b(0)$ are the S matrices for the energy level 'a' (the upper state) and the energy level 'b' (the lower state) involved in the radiative transition, respectively; the symbol $\{\cdots\}$ signifies averaging over the angular variables of vectors $\boldsymbol{\rho}$ and $\mathbf{v}$.

The conventional theory subdivides the collisions into so-called weak collisions (treated by Dirac's standard perturbation theory) and strong collisions. In terms of the impact parameter, the boundary between the two types of collision is determined from the condition in which the expression $\{S_a S_b^* - 1\}$ (formally calculated for the weak collisions) begins to violate the unitarity of the S matrices. This boundary value of the impact parameter is called the electron Weisskopf radius ρ_{We}.

In the conventional theory, the integral over the impact parameters diverges at small values of ρ, which is why the conventional theory divides the range of the integration into two subranges: from zero to ρ_{We} (the strong collisional contribution) and from ρ_{We} to ρ_{max} (the weak collisional contribution). The quantity ρ_{max} is the upper cutoff, required because the integral over the impact parameters also diverges at large ρ. The typical choice of the upper cutoff is the Debye radius $\rho_D = [T/(4\pi e^2 N_e)]^{1/2}$, T being the electron temperature.

In the conventional theory, after the calculation of the S matrices is performed for the weak collisions, the operator of the Stark broadening by plasma electrons is as follows (in atomic units):

$$\Phi_{ab}^{weak} \equiv C \int_{\rho_{max}}^{\rho_{we}} d\rho\, \rho \sin^2 \frac{\Theta(\rho)}{2} = \frac{C}{2} \int_{\Theta_{max}}^{\Theta_{min}} d\Theta \frac{d\rho^2}{d\Theta} \sin^2 \frac{\Theta}{2}, \tag{7.23}$$

where Θ is the scattering angle that describes the outcome of the collision between the perturbing electron and the radiator. The quantities Θ and ρ are interrelated, as elucidated below. In equation (7.23), the operator C has the form:

$$C = -\frac{4\pi}{3} N_e \left[\int_{\infty}^{0} dv\, v^3 f(v) \right] \frac{m^2}{(Z-1)^2} (r_a - r_b{}^*)^2. \tag{7.24}$$

In equation (7.24), m is the reduced mass of the 'perturbing electron—radiator' system, and $f(v)$ is the velocity distribution of the perturbing electrons. Further, in equation (7.24), $\mathbf{r}$ is the operator of the radius vector of the atomic electron (it is proportional to $1/Z$); the subscripts 'a' and 'b' relate to the action of this operator on the upper and lower energy levels, respectively. The formula $(\mathbf{r}_a - \mathbf{r}_b{}^*)^2$ signifies the scalar product (sometimes called the dot product) of the operator $(\mathbf{r}_a - \mathbf{r}_b{}^*)$ with itself. We note that the calculations of the matrix elements of this operator in the theory of the electron-induced Stark broadening in plasmas are performed on the basis of the unperturbed wave functions.

Since, in the conventional theory, the scattering takes place in the effective coulomb potential (in the effective nuclear charge $Z - 1$), the path of the perturbing electron has a hyperbolic shape characterized by the following relation between the scattering angle and the impact parameter:

$$\rho^{(0)} = \frac{Z-1}{m\, v^2} \cot \frac{\Theta}{2}. \tag{7.25}$$

In equation (7.25), the superscript (0) in the impact parameter $\rho^{(0)}$ refers to the hyperbolic paths.

In [20], we analyzed the real situation: the paths of the perturbing electrons are more complex—the nucleus, the bound electron, and the perturbing electron should be treated as a three-body (rather than a two-body) system. Based on the fact that the frequency Ω_{ab} of the spectral line is much greater than the typical frequency v_{Te}/ρ_{We} of the variation in the perturbing electron electric field at the location of the radiator, we employed the analytical method of separating the slow and rapid subsystems. (Details are presented in appendix C.) We note that the frequency Ω_{ab} is the Kepler frequency or its harmonics if radiative transition occurs between the Rydberg states.

Step one of this method is to 'freeze' the perturbing electron (the slow subsystem) and obtain the analytical solution for the radiator (the rapid subsystem); the solution depends on the frozen coordinates of the perturbing electron: namely, on the separation R between the radiator and the perturbing electron. The corresponding energy, within the first nonvanishing order of the dependence on R, is the following:

$$E_{nq}(R) = -\frac{Z^2}{n^2} + \frac{3\,n\,q}{2\,Z\,R^2}, \tag{7.26}$$

where n is the principal quantum number and $q = n_1 - n_2$ is the difference between the parabolic quantum numbers of the radiator state.

Step two is to analyze the motion of the perturbing electron (the slow subsystem) in the 'effective potential' $V_{\text{eff}}(R)$ that combines the actual potential and $E_{nq}(R)$. The effective potential for the motion of the perturbing electron can be represented in the form (given that the first term on the right-hand side of equation (7.26) does not depend on R and thus does not influence the motion):

$$V_{\text{eff}}(R) = -\frac{\alpha}{R} + \frac{\beta}{R^2}, \quad \alpha = Z - 1. \tag{7.27}$$

For the Lyman lines, since the lower state b can be considered to be unperturbed (up to the order $\sim 1/R^2$), the coefficient β can be represented in the form:

$$\beta = \frac{3\,n_{\text{a}}q_{\text{a}}}{2\,Z}. \tag{7.28}$$

For other (non-Lyman) lines of hydrogenlike ions, in order to allow for both the upper and lower states, the coefficient β can be formulated as follows:

$$\beta = \frac{3\,(n_{\text{a}}q_{\text{a}} - n_{\text{b}}q_{\text{b}})}{2\,Z}. \tag{7.29}$$

For the motion in the potential given by equation (7.27), there is an exact analytical solution. In particular, for the interrelation between the impact parameter and the scattering angle, this exact analytical solution yields the following (instead of equation (7.25); see, e.g. book [22]):

$$\Theta = \pi - \frac{2}{\sqrt{1 + \frac{2\,m\,\beta}{M^2}}} \arctan\sqrt{\frac{4\,E}{\alpha^2}\left(\beta + \frac{M^2}{2m}\right)}. \tag{7.30}$$

In equation (7.30), M and E are the perturbing electron angular momentum and the energy, respectively.

In terms of the impact parameter ρ, the angular momentum is

$$M = m\,v\,\rho. \tag{7.31}$$

Equation (7.30) can now be represented in the form:

$$\tan\left(\frac{\pi - \Theta}{2}\sqrt{1 + \frac{2\,\beta}{m\,v^2\rho^2}}\right) = \frac{v}{\alpha}\sqrt{m^2v^2\rho^2 + 2\,m\,\beta}. \tag{7.32}$$

After finding the solution of equation (7.32) and the subsequent substitution of the solution in equation (7.23), one can obtain a more rigorous formula for the operator of the dynamical Stark broadening by electrons compared to that given by the conventional theory. However, this can be done only numerically because equation (7.32) does not allow an exact analytical solution.

In [20], we gave an approximate analytical solution of equation (7.32), as follows. We expanded equation (7.32) in powers of β and obtained the following:

$$\tan\left(\frac{\pi - \Theta}{2}\right) + \left(\frac{\pi - \Theta}{2}\right)\left[1 + \tan^2\left(\frac{\pi - \Theta}{2}\right)\right]\frac{\beta}{m\, v^2\rho^2} \approx \frac{m\, v^2\rho}{\alpha} + \frac{\beta}{\alpha\, \rho}. \quad (7.33)$$

We sought the solution with respect to ρ in the form

$$\rho \approx \rho^{(0)} + \rho^{(1)}. \quad (7.34)$$

In equation (7.34), $\rho^{(1)} \ll \rho^{(0)}$, where $\rho^{(0)}$ corresponds to $\beta = 0$ (and was presented in equation (7.4)). On substituting equation (7.34) into equation (7.33), the result is

$$\frac{(\pi - \Theta)\, \beta}{2\, m\, v^2 \rho^{(0)^2} \sin^2 \dfrac{\Theta}{2}} - \frac{\beta}{\alpha\, \rho^{(0)}} \approx \frac{m\, v^2 \rho^{(1)}}{\alpha}. \quad (7.35)$$

After finding $\rho^{(1)}$ using equation (7.35), the formula for ρ becomes the following:

$$\rho \approx \frac{\alpha}{m\, v^2} \cot \frac{\Theta}{2} + \frac{\beta}{\alpha}\left(\frac{\pi - \Theta}{2 \cos^2 \dfrac{\Theta}{2}} - \tan \frac{\Theta}{2}\right). \quad (7.36)$$

The subsequent integration in equation (7.22) can be simplified if, instead of the integral over ρ, the integration is performed over Θ. To achieve this task, we calculate the square of equation (7.36):

$$\rho^2 \approx \frac{\alpha^2}{m^2\, v^4} \cot^2 \frac{\Theta}{2} + \frac{\beta}{m\, v^2}\left(\frac{\pi - \Theta}{\sin \dfrac{\Theta}{2} \cos \dfrac{\Theta}{2}} - 1\right), \quad (7.37)$$

where we kept only the first-order terms with respect to β. For a further simplification of the formulas, we let $\phi = \Theta/2$. We then differentiated equation (7.37) with respect to ϕ and obtained the following:

$$\begin{aligned}\frac{d\rho^2}{d\phi} \approx &- \frac{\alpha^2}{m^2\, v^4} \frac{2 \cot \phi}{\sin^2\phi} - \frac{2\, \beta}{m\, v^2}\left[\left(\frac{1}{\sin \phi \cos \phi}\right)\right.\\ &\left. + \left(\frac{\pi}{2} - \phi\right)\left(\frac{1}{\sin^2\phi} - \frac{1}{\cos^2\phi}\right)\right].\end{aligned} \quad (7.38)$$

On substituting $\Theta = 2\phi$ and then $\frac{d\rho^2}{d\phi}$ from equation (7.38) into the far right of equation (7.23), we got

$$\Phi_{ab}^{weak} = -C\left[\frac{\alpha^2}{m^2v^4}\int_{\phi_{max}}^{\phi_{min}} \cot\phi\, d\phi + \frac{\beta}{m\,v^2}\int_{\frac{\pi}{2}}^{0} \tan\phi\, d\phi + \frac{\beta}{m\,v^2}\int_{\frac{\pi}{2}}^{0}\left(\frac{\pi}{2} - \phi\right)(1 - \tan^2\phi)\, d\phi\right]. \tag{7.39}$$

After performing the integrations in equation (7.39), we obtained the following expression for the weak collisional contribution to the operator of the dynamical Stark broadening by plasma electrons:

$$\Phi_{ab}^{weak} = -\frac{4\pi}{3}N_e(\boldsymbol{r}_a - \boldsymbol{r}_b{}^*)^2\left[\int_{\infty}^{0} dv\,\frac{f(v)}{v}\right]\left[\log\frac{\sin\phi_{max}}{\sin\phi_{min}} + \frac{mv^2\beta}{(Z-1)^2}\left(\frac{\pi^2}{4} - 1\right)\right]. \tag{7.40}$$

We then added the following formula for the contribution of strong collisions:

$$\Phi_{ab}^{strong} \approx \pi v N_e \rho_W e^2, \tag{7.41}$$

where ρ_{We} corresponds to ϕ_{max}. The formulas for ϕ_{max} and ϕ_{min} from [16] are as follows:

$$\sin\phi_{max} = \sqrt{\frac{3}{2}}\,\frac{Z(Z-1)}{(n_a^2 - n_b^2)m\,v}, \tag{7.42}$$

$$\sin\phi_{min} = \frac{\dfrac{Z-1}{m\,v^2\rho_D}}{\sqrt{1 + \dfrac{(Z-1)^2}{m^2v^4\rho_D^2}}}. \tag{7.43}$$

It should be underscored that the formula from [16] reproduced here as equation (7.42) was an approximate expression for non-Lyman lines. For Lyman lines, the lower level (the ground level) does not yield any contribution to the operator of the dynamical Stark broadening by plasma electrons. Therefore, for the Lyman lines, we can write the following instead of equation (7.42):

$$\sin\phi_{max} = \sqrt{\frac{3}{2}}\,\frac{Z(Z-1)}{n_a^2 m\,v}. \tag{7.44}$$

It should be noted that the right-hand sides of equations (7.42) or (7.44) can become greater than unity for relatively slow perturbing electrons. In this situation, one should use $\sin\phi_{max} = 1$. This corresponds to $\rho_{min} = 0$ and thus a zero strong collisional contribution. Usually, the statistical weight of such slow electrons in the velocity distribution of plasma electrons is relatively small.

On substituting the expressions for $\sin\phi_{max}$ and $\sin\phi_{min}$ into equation (7.39) and adding together the weak and strong collisional contributions, we found the following final result for the operator of the dynamical Stark broadening by plasma electrons:

$$\Phi_{ab}(\beta) = -\frac{4\pi}{3}N_e(\boldsymbol{r}_a - \boldsymbol{r}_b{}^*)^2\left[\int_\infty^0 dv\,\frac{f(v)}{v}\right]\left\{\frac{1}{2}\left[1 - \frac{3}{2}\frac{Z^2(Z-1)^2}{(n_a^2 - n_b^2)^2 m^2 v^2}\right]\right.$$
$$\left. + \log\left[\sqrt{\frac{3}{2}}\frac{Z\,v\,\rho_D}{(n_a^2 - n_b^2)}\sqrt{1 + \left(\frac{Z-1}{m\,v^2\rho_D}\right)^2}\right] + \frac{mv^2\beta}{(Z-1)^2}\left(\frac{\pi^2}{4} - 1\right)\right\} \tag{7.45}$$

for the non-Lyman lines and

$$\Phi_{ab}(\beta) = -\frac{4\pi}{3}N_e(\boldsymbol{r}_a - \boldsymbol{r}_b{}^*)^2\left[\int_\infty^0 dv\,\frac{f(v)}{v}\right]\left\{\frac{1}{2}\left[1 - \frac{3}{2}\frac{Z^2(Z-1)^2}{n_a^4 m^2 v^2}\right]\right.$$
$$\left. + \log\left[\sqrt{\frac{3}{2}}\frac{Z\,v\,\rho_D}{n_a^2}\sqrt{1 + \left(\frac{Z-1}{m\,v^2\rho_D}\right)^2}\right] + \frac{mv^2\beta}{(Z-1)^2}\left(\frac{\pi^2}{4} - 1\right)\right\} \tag{7.46}$$

for the Lyman lines. In equations (7.45) and (7.46), the symbol log[···] signifies the natural logarithm.

To reveal the importance of allowing for the non-hyperbolic trajectories, we analyzed the ratio of the third term in braces in equation (7.46) to the combination of the first and second terms:

$$\text{ratio} = \frac{\frac{3}{2}\frac{mv^2(n_a q_a - n_b q_b)}{(Z-1)^2}\left(\frac{\pi^2}{4} - 1\right)}{\frac{1}{2}\left[1 - \frac{3}{2}\frac{Z^2(Z-1)^2}{(n_a^2 - n_b^2)^2 m^2 v^2}\right] + \log\left[\sqrt{\frac{3}{2}}\frac{Z\,v\,\rho_D}{(n_a^2 - n_b^2)}\sqrt{1 + \left(\frac{Z-1}{m\,v^2\rho_D}\right)^2}\right]} \tag{7.47}$$

for the non-Lyman lines or the ratio

$$\text{ratio} = \frac{\frac{3}{2}\frac{mv^2 n_a q_a}{(Z-1)^2}\left(\frac{\pi^2}{4} - 1\right)}{\frac{1}{2}\left[1 - \frac{3}{2}\frac{Z^2(Z-1)^2}{n_a^4 m^2 v^2}\right] + \log\left[\sqrt{\frac{3}{2}}\frac{Z\,v\,\rho_D}{n_a^2}\sqrt{1 + \left(\frac{Z-1}{m\,v^2\rho_D}\right)^2}\right]} \tag{7.48}$$

for the Lyman lines. We provided a quantitative example for some Lyman lines. Instead of integrating over the velocities of the perturbing electrons, we utilized the mean thermal velocity v_T, which is usually used in the theories of the dynamical Stark broadening. Below are the formulas for v_T, the Debye radius ρ_D, and the reduced mass, m:

$$v_T = 0.1917\sqrt{\frac{T(\text{eV})}{m}}, \quad \rho_D = 1.404 \times 10^{11}\sqrt{\frac{T(\text{eV})}{N_e(\text{cm}^{-3})}}, \quad m = \frac{1 + \dfrac{m_e}{A\, m_p}}{1 + \dfrac{2\, m_e}{A\, m_p}}, \tag{7.49}$$

where A is the atomic number of the radiator ($A \approx 2Z$), m_p is the proton mass, and m_e is the electron mass.

As examples, for several Lyman lines of ionized helium, table 7.1 provides the numerical values of the ratio from equation (7.48) for a plasma that has the following parameters: $N_e = 2 \times 10^{17}$ cm^{-3}, $T = 8$ eV.

Figure 7.6 demonstrates the dependence of the ratio given by equation (7.48) on the electron density N_e at the temperature $T = 8$ eV for the Stark components characterized by the quantum number $|q| = 1$ for the following spectral lines: Lyman-alpha ($n = 2$), Lyman-beta ($n = 3$), and Lyman-gamma ($n = 4$).

From table 7.1 and figure 7.6, one can see that the allowance for the non-hyperbolic trajectories of the perturbing electrons already becomes important for the Lyman lines of the ionized helium at $N_e \sim 10^{17}$ cm^{-3}. The effect becomes more and more important as the electron density increases.

We note in passing that in the case in which the formally calculated ratio given by equation (7.48) is comparable to unity, this would mean that the approximate analytical solution is invalid. In this case, a numerical solution of equation (7.32) with respect to ρ should be used.

In relation to the nonlinearity, we wish to emphasize the following. The ratio given by equation (7.48), presented in figure 7.6, shows the nonlinear dependence on the electron density N_e. As for the dependence of the Stark width on N_e, it is proportional to the right-hand sides of equations (7.45) or (7.46). After introducing

Table 7.1. Ratios given by equation (7.26) for the Stark components of several Lyman lines of He II at a temperature $T = 8$ eV and an electron density $N_e = 2 \times 10^{17}$ cm^{-3}. Reproduced from [20].

n	$\lvert q \rvert$	Ratio
2	1	0.3261
3	1	0.3748
3	2	0.7496
4	1	0.5156
4	2	1.0311
4	3	1.5467

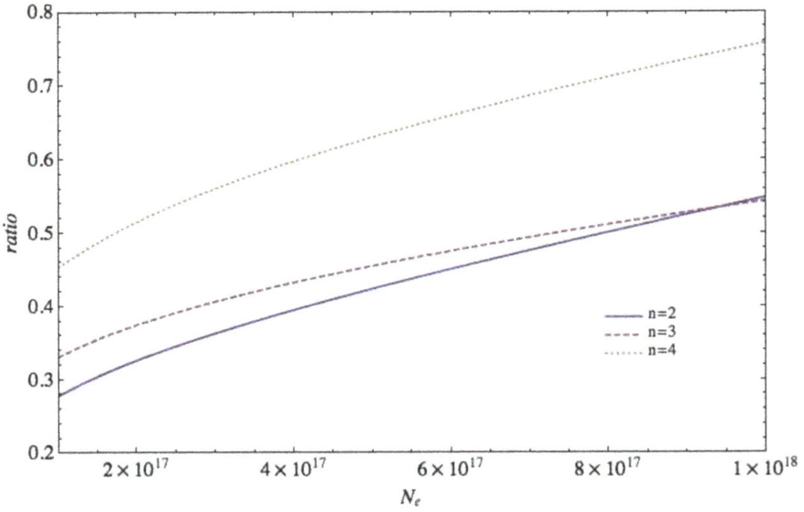

Figure 7.6. The dependence of the ratio given by equation (7.48) on the electron density N_e at the temperature $T = 8$ eV for the Stark components characterized by the quantum number $|q| = 1$ for the following spectral lines: Lyman-alpha ($n = 2$), Lyman-beta ($n = 3$), and Lyman-gamma ($n = 4$). Reproduced from [20].

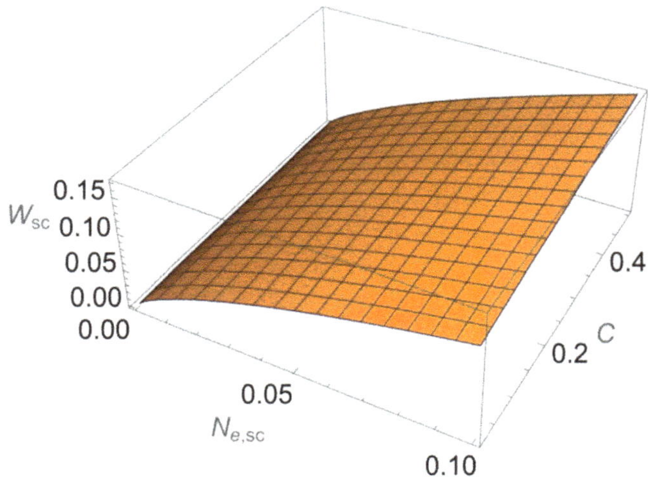

Figure 7.7. The dependence of the scaled electron-induced dynamical Stark width W_{sc} on the scaled electron density $N_{e,sc}$ and on the parameter C from equation (7.50).

the properly scaled electron density $N_{e,sc}$ and the dynamical Stark width W_{sc}, the dependence of the latter on $N_{e,sc}$ can be represented in the following form:

$$W_{sc} = N_{e,sc}\left\{C + \ln\left[(1 + 1/N_{e,sc})^{1/2}\right]\right\}. \tag{7.50}$$

In equation (7.50), the quantity C depends on the temperature T, the nuclear charge Z, and quantum numbers but is independent of $N_{e,sc}$.

Figure 7.7 shows the dependence of the scaled electron-induced dynamical Stark width W_{sc} on the scaled electron density $N_{e,sc}$ and on the parameter C.

Figure 7.8 represents the same plot as in figure 7.7, but from the alternative viewpoint, so that together with figure 7.7 it shows the dependence of W_{sc} on $N_{e,sc}$ in a comprehensive way.

Figure 7.9 presents the dependence of W_{sc} on $N_{e,sc}$ for the following three values of the parameter C: 0.02, 0.1, and 0.5.

From figures 7.7–7.9, it can be seen that the dependence of the electron-induced dynamical Stark width on the electron density is nonlinear.

It should be underscored that the possibility of obtaining the above analytical solution was mainly facilitated by the algebraic (higher than geometrical) symmetry of the potentials $V(R) = -A/R + B/R^2$, where A and B are constants. The algebraic symmetry is manifested by the existence of an additional conserved quantity

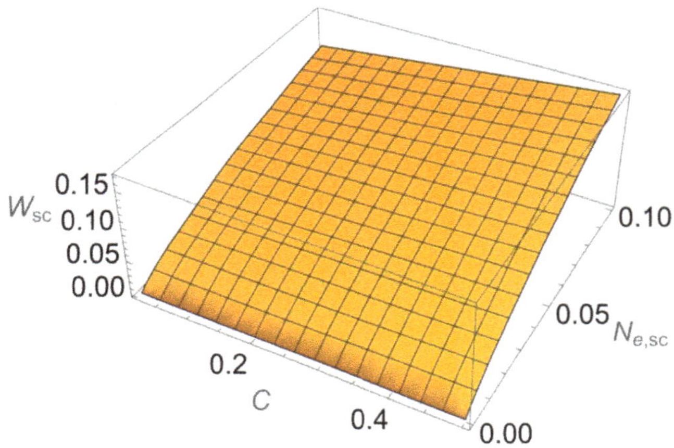

Figure 7.8. The same plot as in figure 7.7, but from the alternative viewpoint, so that together with figure 7.7 it shows the dependence of W_{sc} on $N_{e,sc}$ in a comprehensive way.

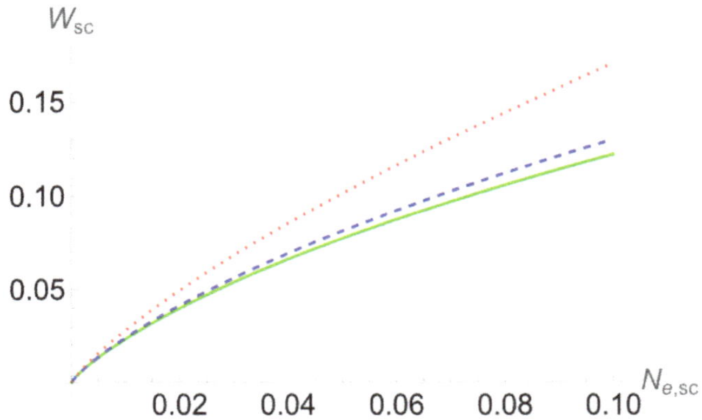

Figure 7.9. Dependence of W_{sc} on $N_{e,sc}$ for the following three values of the parameter C from equation (7.50): 0.02 (solid line), 0.1 (dashed line), and 0.5 (dotted line).

$M_{eff}^2 = M^2 + 2mB$, where m and M are the mass and the angular momentum of the particle, respectively. This conserved quantity can be treated as the square of the *effective* angular momentum.

As for the utilization of the impact approximation, it was not critical to our study. Rather, we employed it to simplify the formulas and thus the message of our work, as well as for a comparison with the corresponding results produced by the conventional theory, for which the impact approximation was critical. A brief description of the impact approximation is presented in appendix D.

Finally, we mention [23], which contains noteworthy results on the same subject (presented also in [24]). The results were obtained in the framework on the so-called quantum theory of the dynamical Stark broadening, in which both the perturbing plasma electrons and the radiator are treated on an equal footing by quantum mechanics. In distinction, our results were produced in the framework of the corresponding semiclassical theory, in which the perturbing plasma electrons are described by classical mechanics whereas the radiator is described by quantum mechanics. (The same semiclassical framework was used in the conventional theory of the dynamical Stark broadening.) Both in our work [20] and in [23], the symmetry of the potentials $V(R) = -A/R + B/R^2$ was used to produce analytical results.

The authors of [23] employed the impact approximation to produce specific results for the Stark width of Lyman lines in the classical limit only. In [20], we also compared their results with the corresponding results of the conventional theory for ionized helium Lyman lines for $N_e \sim (10^{17}–10^{18})$ cm^{-3}. This is the range of N_e used by most of the experiments that have measured the widths of the ionized helium lines. We found that the Stark widths given in [23] are two or more orders of magnitude greater than the corresponding Stark widths calculated by the conventional theory.

It should be emphasized that the widths of the ionized helium lines measured in experiments where plasma parameters are determined independently of the width of these spectral lines (i.e. in the benchmark experiments) always show the ratio of the experimental width to the corresponding width given by the conventional theory to be a factor of two or less. The same is true for the corrections in our paper [20] to the Stark width calculated by the conventional theory.

7.4 The generalized theory

The generalized theory of the dynamical Stark broadening of hydrogenic spectral lines in plasmas was developed in [25–27]. This theory is based mostly on a *generalization* of the formalism of dressed atomic states (DAS) in plasmas.

The usual formalism of dressed atomic states was initially intended for describing the interaction of a *monochromatic* (or quasi-monochromatic) field (such as the electric field of laser or maser radiation) with gases. Subsequently, this formalism was employed to describe the interaction of the quasi-monochromatic electric field with plasmas [28]. The utilization of this formalism in plasma spectroscopy facilitated better accuracy of the analytical calculations and more advanced codes.

In contrast, in the *generalization* of the dressed atomic-state formalism in [25–27], the atomic states were dressed by a *broadband* field of plasma electrons and ions. Obviously, these generalized dressed atomic states result in a more complicated construct than the usual dressed atomic states.

The design of the generalized theory consists of the following eight stages, as formulated in [27], from which we quote these stages here:

1. *Separation of static and dynamic ions.* For a given value τ of the argument of the correlation function $C(\tau)$, the ion-dynamical part of $C(\tau)$ originates from the collisions for which the instants of the closest approach t_0 fall within the interval $(-\tau/2,\tau/2)$. The rest of the perturbing ions are considered to be static. This is consistent with the fact that at $\tau \to \infty$, all ions are dynamic, while at $\tau \to 0$, all ions are static. Thus, different values of τ correspond different proportions of dynamic and static ions.
2. *Partition of the Hamiltonian.* The Hamiltonian $H(t)$ is broken down into the following four terms:

$$H(t) = H_0 + V^{IS} + V_{par}(t) + V_{perp}(t),\ V_{par}(t) = V_{||}^{ID}(t) + V_{||}^{E}(t),\ V_{perp}(t) = V_{\perp}^{ID}(t) + V_{\perp}^{E}(t). \tag{7.51}$$

Here, H_0 is the Hamiltonian of the unperturbed radiating atom/ion (radiator); the superscripts IS, ID, and E refer to the interactions of the radiator with static ions, dynamic ions, and electrons, respectively; the subscripts || and $\perp$ refer, respectively, to parallel and perpendicular components of the electron and dynamic ion microfields with respect to the direction of the quasistatic ion field at the location of the radiator.

3. *Generalized dressed atomic states.* The atomic states of the radiator are dressed by the following time-dependent factor:

$$Q(t) = \exp\left\{(i/\hbar)\left[(H_0 + V^{IS})t + \int_{-\infty}^{t} dt' V_{par}(t')\right]\right\}. \tag{7.52}$$

In other words, the interactions V^{IS} and $V_{par}(t')$ are taken into account *exactly*, moreover—*analytically*. The interaction $V_{perp}(t)$ is then allowed for by Dirac's perturbation theory.

The dressing factor $Q(t)$ is controlled by the interaction $V_{par}(t)$ with a *broadband* field—in distinction to the usual DAS dressed by a monochromatic field (examples of the latter can be found in book [28]).

4. *Evolution operator in the MODIFIED interaction representation.* This employs the dressing factor $Q(t)$ from equation (7.52):

$$U(t_1, t_2) = \mathrm{T}\exp\left[-(i/\hbar)\int_{t_1}^{t_2} dt Q^*(t) V_{perp}(t) Q(t)\right], \tag{7.53}$$

where Texp[$\cdots$] is the time-ordered exponential.

5. *Calculation of the M-matrix.* The *M*-matrix that controls the calculation of the correlation function (defined in stage 6) is calculated as follows:

$$M(t_1, t_2) = \exp\left[(\mathrm{i}/\hbar)\int_{t_1}^{t_2} \mathrm{d}t' V_{\mathrm{par}}(t')\right] U(t_1, t_2), \tag{7.54}$$

where exp[⋯] is calculated *exactly*, analytically, while $U(t_1,t_2)$ is calculated via the perturbation theory. The impact approximation is *not* used.

6. *Correlation function.* This is defined by the formula

$$C(\tau) = \mathrm{Tr}\,\{\mathbf{dd}\exp[-(\mathrm{i}/\hbar)(H_0 + V^{\mathrm{IS}})\tau]\rho\langle M(\tau/2, -\tau/2)\rangle_{\mathrm{av}}\}, \tag{7.55}$$

where Tr{⋯} stands for the trace, ⟨⋯⟩av denotes the ensemble average.

7. *Line shape at a fixed quasistatic ion field F.* This is calculated as the Fourier transform of the correlation function $C(\tau)$.
8. *Averaging over the distribution W(F).* The distribution $W(F)$ of the quasi-static ion field can be produced using, e.g. APEX code [29].'

The central point of the generalized theory is the following: *it treats the interaction $V_{\|}(t)$ on an equal footing with the unperturbed Hamiltonian H_0: no perturbation expansion is involved.*

In order to compare the generalized theory with the conventional theory, it is useful to introduce the following parameter that plays a central role in the generalized theory:

$$Y \equiv \frac{<\mathrm{V}^{\mathrm{IS}}>}{\Omega_W(\mathrm{V}_{\mathrm{par}})}, \tag{7.56}$$

where $\Omega_{\mathrm{W}}(V_{\mathrm{par}})$ is the Weisskopf frequency related to the dynamical interaction $V_{\mathrm{par}}(t)$. (The Weisskopf frequency is defined as the mean thermal velocity of perturbing charges divided by the Weisskopf radius.) The generalized theory embraces the conventional theory as the limiting case corresponding to $Y \to 0$.

In the generalized theory, the integral over the impact parameters *converges* at small impact parameters, while in the conventional theory for neutral radiators, the corresponding integral diverges at small impact parameters. The physical reason for the convergence of this integral in the generalized theory is as follows. The interaction $V_{\mathrm{perp}}(t)$ causes the virtual transitions between the sublevels of the radiator that are *'dressed'* by the interaction $V_{\mathrm{par}}(t)$. Due to this dressing, the divergence of the integral is eliminated and the accuracy of the results is enhanced.

The matrix $M(t_2,t_1)$ in equation (7.54) and its simplified counterpart—the scattering matrix $S = M(-\infty, \infty)$—are composed of two terms that differ physically:

$$M = M_{\mathrm{a}} + M_{\mathrm{na}}, \tag{7.57}$$

$$\hat{S} = \hat{S}_a + \hat{S}_{na}. \tag{7.58}$$

The first term in equations (7.57) and (7.58) is the *adiabatic* contribution. This is analogous to the corresponding concept known as the 'Old Adiabatic Theory' of broadening (see e.g. review [30]). For example, the form of the adiabatic term in the scattering matrix is as follows:

$$S_a \equiv exp\left[\frac{i}{\hbar}\int_{\infty}^{\infty} dt\hat{V}_1(t)\right]. \tag{7.59}$$

Here and below, it is denoted by $V_1 = V_{\text{par}}$, $V_2 = V_{\text{perp}}$.

The second term in equations (7.57) and (7.58) is nonadiabatic: it is nonzero only because of the nonadiabatic virtual transitions produced by the interaction $V_2(t)$ between sublevels of the radiator dressed by the interaction $V_1(t)$. The nonadiabatic term in the scattering matrix is as follows:

$$\hat{S}_{na} \equiv exp\left[\frac{i}{\cdot}\int_{\infty}^{\infty} dt\hat{V}_1(t)\right]\left\{\hat{T}\, exp\left[\frac{i}{\hbar}\int_{\infty}^{\infty} dt\hat{Q}^*\hat{V}_2(t)\hat{Q}\right]\right\}, \tag{7.60}$$

where the first nonvanishing contribution, obtained after expanding Texp[···] and performing the angular averaging, is typically of the second order with respect to V_2.

It should be emphasized that the generalized theory, despite having more complex starting expressions than those of the conventional theory, was *analytically* developed to the same extent as the conventional theory. We give some details below.

For a hydrogen/deuterium atom subjected to the quasistatic part **F** of the ion microfield and the dynamic field **E**(t) of the plasma electrons, the Hamiltonian is

$$H = H_0 - \mathbf{dF} - \mathbf{dE}(t). \tag{7.61}$$

In equation (7.61), H_0 is the Hamiltonian in the absence of the external fields and **d** is the operator of the electric dipole moment. The choice of the z-axis of the parabolic quantization to be along the quasistatic field **F** diagonalizes the operator –**dF** in the n-subspace, corresponding to a fixed value of the principal quantum number n. In the version of the conventional theory developed in paper [31], the interaction –**dF** was taken into account 'exactly' (within the n-subspace). The interaction –**dE**(t) was processed in the second order of Dirac's perturbation theory.

In contrast, the generalized theory allowed exactly (within the n-subspace) for the entire z-component of the total field **F** + **E**(t). This was possible because not only $-d_zF$, but also $-d_z[F + E_z(t)]$ is diagonal in any n-subspace. In this way, the z-component of the electron microfield was taken into account more rigorously than in the conventional theory.

The perturbed wave functions are set up as follows:

$$\psi_j(t, \mathbf{r}) = \exp\left[-\mathrm{i}\omega_j t + \mathrm{i}\int_{-\infty}^{t} \mathrm{d}t'(d_z)_{jj}E_z(t')\right] U(t, -\infty)\varphi_j(\mathbf{r}) \\ , \omega_j = \omega_{0j} - (d_z)_{jj}F. \tag{7.62}$$

In equation (7.62), ω_{0j} is the (unperturbed) energy of the Stark sublevel j ($j = \beta, \beta', \ldots$ for the lower sublevels involved in the radiative transition or $j = \alpha, \alpha', \ldots$ for the upper sublevels involved in the radiative transition). Here and below, we set $\hbar = 1$.

In our modified interaction representation, the evolution operator $U(t_2,t_1)$ is

$$U(t_2, t_1) = \mathrm{T}\exp\left[-\mathrm{i}\int_{t_1}^{t_2} \mathrm{d}t\ V(t)\right], \tag{7.63}$$

$$V(t) = \exp\left[\mathrm{i}\int_{-\infty}^{t} \mathrm{d}t'\ H_1(t')\right][-d_{\perp}E_{\perp}(t)]\exp\left[-\mathrm{i}\int_{-\infty}^{t} \mathrm{d}t'\ H_1(t')\right]. \tag{7.64}$$

In equation (7.64), $d_{\perp}E_{\perp} = d_xE_x + d_yE_y$. The operator $H_1(t)$ is

$$H_1(t) = H_0 - d_zF - d_zE_z(t). \tag{7.65}$$

In the expression for the shape of the spectral line,

$$J(\Delta\omega) = (2\pi)^{-1}\int_{-\infty}^{\infty} \mathrm{d}\tau C(\tau)\exp(\mathrm{i}\Delta\omega\tau), \tag{7.66}$$

where $\Delta\omega = \omega - \omega_0$, the correlation function can be written as

$$C(\tau) = \sum_{\alpha,\, \alpha',\, \beta,\, \beta'} \langle a \mid \mathbf{d} \mid \beta'\rangle\langle\beta \mid \mathbf{d} \mid \alpha'\rangle \exp[-\mathrm{i}(\omega_\alpha - \omega_\beta)\tau]\langle\alpha'|\langle\beta'|[M(\tau/2, -\tau/2)]_{\mathrm{av}}|\beta\rangle|\alpha\rangle. \tag{7.67}$$

In equation (7.67), the relation of the M-operator to the evolution operator (7.63) is

$$M(\tau/2, -\tau/2) = \exp\left[\mathrm{i}d_z\int_{-\tau/2}^{\tau/2} dtE_z(t)\right]U(\tau/2, -\tau/2). \tag{7.68}$$

Since, in the M-operator, there are adiabatic and nonadiabatic contributions, the same is true for the correlation function $C(\tau)$ (see equation (7.67)). The adiabatic term in the correlation function, which is proportional to d_zd_z, is diagonal in the line space. The other term in the correlation function, controlled by $d_xd_x + d_yd_y$, is not diagonal in the line space.

The diagonal contribution to the correlation function has the form

$$\begin{aligned} C_{\mathrm{diag}}(\tau) = {} & |\mathbf{d}_{\alpha\beta}|^2 \exp[-i(\omega_\alpha - \omega_\beta)\tau] \\ & \left\{\exp\left[\mathrm{i}\Delta d_z\int_{\tau/2}^{\tau/2} \mathrm{d}tE_z(t)\right] U_{\alpha\alpha}(\tau/2, -\tau/2)U_{\beta\beta}{}^*(\tau/2, -\tau/2)\right\}_{\mathrm{av}}. \end{aligned} \tag{7.69}$$

In equation (7.69),

$$\Delta d_z \equiv (d_z)_{\alpha\alpha} - (d_z)_{\beta\beta}. \tag{7.70}$$

On expanding the evolution operator in equation (7.69) up to the second order, we get

$$C_{\mathrm{diag}}(\tau) = |\mathbf{d}_{\alpha\beta}|^2 \exp[-i(\omega_\alpha - \omega_\beta)\tau][\langle f_{\mathrm{a}}(\tau)\rangle_{\mathrm{av}} + \langle f_{\mathrm{na}}(\tau)\rangle_{\mathrm{av}}]. \tag{7.71}$$

In equation (7.71), the adiabatic part is controlled by

$$f_{\mathrm{a}}(\tau) \equiv \exp\left[\mathrm{i}\Delta d_z\int_{-\tau/2}^{\tau/2} \mathrm{d}tE_z(t)\right], \tag{7.72}$$

while the nonadiabatic part is controlled by

$$f_{\text{na}}(\tau) \equiv -f_{\text{a}}(\tau)\left\{\sum_{\alpha'}\int_{-\tau/2}^{\tau/2}\mathrm{d}t_1\int_{-\tau/2}^{\tau/2}\mathrm{d}t_2 \exp\left[\mathrm{i}(\omega_\alpha - \omega_{\alpha'})(t_1 - t_2) - \mathrm{i}(\delta d_z)_\alpha \int_{t_2}^{t_1}\mathrm{d}t\mathrm{d}t\, E_z(t)E_z(t)\right][(d_\perp)_{\alpha\alpha'}E_\perp(t_1)][(d_\perp)_{\alpha'\alpha}E_\perp(t_2)] + \sum_{\beta'}[\alpha \to \beta,\, \alpha' \to \beta']^*\right\}, \tag{7.73}$$

$$(\delta d_z)_\alpha \equiv (d_z)_{\alpha\alpha} - (d_z)_{\alpha'\alpha'}. \tag{7.74}$$

In equation (7.73), the symbol $\Sigma_{\beta'}$ $[\alpha\to\beta,\ \alpha'\to\beta']$ stands for the term obtained from the term $\Sigma_{\alpha'}[\cdots]$ in equation (7.73) on substituting $\alpha\to\beta$, $\alpha'\to\beta'$.

Equations (7.63)–(7.74) describe the most general form of the generalized theory. This general form was employed for most of the numerical examples presented in [18].

We also present below the simplified generalized theory that uses the impact approximation. The simplified version provides a better insight into the physics behind the generalized theory and facilitates a better understanding of the relation between the generalized and conventional theories.

The limits of the integration in equation (7.68) are now extended to infinities, the corresponding notation being

$$[M(\infty,\, -\infty)]_{\text{av}} \equiv \Phi. \tag{7.75}$$

The operator Φ is called the electron impact operator (also known as the electron broadening operator). Obviously, it does not depend on τ, while still depending on **F**.

The correlation function $C(\tau)$ is now

$$C(\tau) = \sum_{\alpha,\,\alpha',\,\beta,\,\beta'} \langle \alpha \mid \mathbf{d} \mid \beta'\rangle\langle \beta \mid \mathbf{d} \mid \alpha'\rangle \exp[-i(\omega_\alpha - \omega_\beta)\tau]\langle\alpha'|\langle\beta'|\Phi|\beta\rangle|\alpha\rangle, \tag{7.76}$$

and the line shape $J(\Delta\omega)$ is

$$J(\Delta\omega) = -\pi^{-1}\,\mathrm{Re}\left\{\sum_{\alpha,\,\alpha',\,\beta,\,\beta'}\left\langle\alpha|d|\beta'\right\rangle\left\langle\beta|d|\alpha'\right\rangle\left\langle\alpha'\left|\left\langle\beta'\right|[\mathrm{i}(\Delta\omega + \omega_\beta - \omega_\alpha) + \Phi)]^{-1}\left|\beta\right\rangle\right|\alpha\right\rangle\right\}. \tag{7.77}$$

If the operator **dd** were diagonal, then the line shape would consist of Lorentzians of the half width at the half maximum (HWHM) $\Gamma_{\alpha\beta}$ shifted by $D_{\alpha\beta}$, where

$$\Gamma_{\alpha\beta} = -\mathrm{Re}\langle\alpha|\langle\beta|\Phi|\beta\rangle|\alpha\rangle, \quad D_{\alpha\beta} = -\mathrm{Im}\langle\alpha|\langle\beta|\Phi|\beta\rangle|\alpha\rangle. \tag{7.78}$$

However, in fact, the operator **dd** is not diagonal: its nondiagonal elements couple the would-be Lorentzians. This complicates calculations of the line shape, requiring inversion of the matrix of the rank $(n_\alpha n_\beta)^2$, which for the example of the Balmer-gamma line is a matrix of rank 100.

The *adiabatic* part of the electron impact operator has only diagonal elements:

$$\ll \alpha\beta\ \Phi_{a\ d}\ \alpha\ \beta \gg = \frac{4\pi^2 N_e}{3m_e^2 a_0^2} \int_\infty^0 d\,v W_M(v)\,\frac{1}{v}$$
$$\left[\ z_{\alpha\alpha} z_{\alpha\alpha} - 2 z_{\alpha\alpha} z_{\beta\beta} + z_{\beta\beta} z_{\beta\beta}\ \right] I(u_{\alpha\beta}). \tag{7.79}$$

In equation (7.79),

$$u_{\alpha\beta} = m_e v \rho_D / \left[3\hbar X_{\alpha\beta}\right], \quad X_{\alpha\beta} = (n_\alpha q_\alpha - n_\beta q_\beta), \tag{7.80}$$

$$I(u) = \int_0^u \mathrm{d}z\, z[1 - z \sin(1/z)] = \mathrm{Int}(u).$$

The integration can be performed analytically to yield:

$$\mathrm{Int}(u) = u^2/2 - [u^2 \cos(1/u) + (2u^3 - u)\sin(1/u) + Ci(1/u)]/6. \tag{7.81}$$

In equation (7.80), ρ_D is the Debye radius and $q = n_1 - n_2$ is the difference between the parabolic quantum numbers n_1 and n_2. In equation (7.81), $Ci(x)$ is the integral cosine function.

It is very important to underscore that the adiabatic part of the electron impact operator presented in equations (7.79)–(7.81) is equivalent to the combined effect of all orders of the Dyson expansion of the time-ordered exponential in equation (7.63). This is the crucial advantage of the generalized theory over the conventional theory, in which both adiabatic and nonadiabatic terms were calculated only in the first nonvanishing order of the Dyson expansion.

Here is the *nonadiabatic* part of the electron impact operator:

$$\langle\langle \alpha\beta |\ \Phi_{n\ a} |\ \alpha'\ \beta'\ \rangle\rangle = \frac{4\pi\hbar^2 N_e}{3m_e^2 a_0^2} \int_\infty^0 \mathrm{d}v\ W_M(v)\,\frac{1}{v}$$
$$\times \left[\sum_{\alpha''} (x_{\alpha\alpha''} x_{\alpha''\alpha'} + y_{\alpha\alpha''} y_{\alpha''\alpha'}) \int_\infty^0 \frac{d\,Z_\alpha}{Z_\alpha}\, C_\pm(\chi_{\alpha'}, Y_\alpha, Z_\alpha) \right. \tag{7.82}$$
$$\left. + \sum_{\alpha''} [(\alpha\alpha''\alpha'') \to (\beta\beta''\beta'')]^* \right].$$

In equation (7.82), the formula under the summation over β'' can be retrieved from the formula under the summation over α'' by substituting $(\alpha\alpha'\alpha'') \to (\beta\beta'\beta'')$ and performing the complex conjugation. The lower system of the signs relates to the diagonal elements ($\alpha'=\alpha$ in the first term or $\beta'=\beta$ in the second term). The upper system of the signs relates to the nondiagonal elements ($\alpha'\neq\alpha$ or $\beta'\neq\beta$) of the operator **dd**.

Further, in equation (7.82), Z_α and Z_β represent the scaled dimensionless impact parameter:

$$Z_{\alpha,\beta} = 3n_{\alpha,\beta}\hbar F\rho/(2m_e e v). \tag{7.83}$$

In equation (7.83), F is the quasistatic part of the plasma ion microfield.

Below are the formulas for the generalized broadening functions $C_{\pm}$ that appear in equation (7.82):

$$C_{\pm}(\chi, Y, Z) = -\frac{3}{4}\int_{\infty}^{\infty} dx_1 \int_{\infty}^{x_1} dx_2[w(x_1)w(x_2)]^3 \exp[iZ(x_1 \pm x_2)]$$
$$\times \{j_0(\varepsilon) + (2x_1x_2-1)\frac{j_1(\varepsilon)}{\varepsilon} + [(1 - x_1x_2)\sigma_1^2 - (x_1 + x_2)\sigma_1\sigma_2]j_2(\varepsilon)/\varepsilon^2\},$$
$$\varepsilon \equiv \sqrt{\sigma_1^2 + \sigma_2^2},\ \sigma_1 \equiv \frac{Y}{Z}[x_1w(x_1) \pm x_2w(x_2) + 1 \pm 12\chi],$$
$$\sigma_2 \equiv \frac{Y}{Z}[w(x_1) \pm w(x_2)],\ w(x) \equiv 1/\sqrt{1 + x^2}. \tag{7.84}$$

In equation (7.84), $j_0(\varepsilon)$, $j_1(\varepsilon)$, and $j_2(\varepsilon)$ are the spherical Bessel functions:

$$j_0(\varepsilon) = \frac{1}{\varepsilon}\sin\varepsilon,\quad j_1(\varepsilon) = \frac{1}{\varepsilon^2}(\sin\varepsilon - \varepsilon\cos\varepsilon),$$
$$j_2(\varepsilon) = \frac{1}{\varepsilon^3}(3\sin\varepsilon - 3\varepsilon\cos\varepsilon - \varepsilon^2\sin\varepsilon). \tag{7.85}$$

In the nonadiabatic part of the electron impact operator, there are new parameters that were absent from the conventional theory:

$$\chi_a \equiv (n_\alpha q_\alpha \delta_{\alpha\alpha'}\ n_\beta q_\beta \delta_{\beta\beta'})\ /\ (n_\alpha(q_\alpha\ q_{\alpha',})),$$
$$\chi_b \equiv (n_\alpha q_\alpha \delta_{\alpha\alpha'}\ n_\beta q_\beta \delta_{\beta\beta'})\ /\ (n_\beta(q_\beta\ q_{\beta',})). \tag{7.86}$$

$$Y_k = \left(\frac{3n_{k'}}{2Z_k m_e v}\right)^2 \frac{F}{e},\quad k = \alpha, \beta. \tag{7.87}$$

The parameter Y helps the generalized theory to embrace the conventional theory. Indeed, in the limit $Y \to 0$ (meaning that a plasma either has a relatively high temperature, a relatively low density, or both), equation (7.84) simplifies to the corresponding formula of the conventional theory:

$$C_{\pm}(\chi, 0, Z) = -\frac{1}{2}\int_{+\infty}^{-\infty} dx_1 \int_{x_1}^{-\infty} \frac{dx_2\ (1 + x_1 x_2)}{[(1 + x_1^2)(1 + x_2^2)]^{\frac{3}{2}}} exp[iZ(x_1 \pm x_2)]. \tag{7.88}$$

As the parameter Y increases (meaning either a relatively low plasma temperature, or a relatively high electron density, or both), the conventional theory becomes more and more inaccurate.

In the generalized theory, the integral over the scaled impact parameter Z in equation (7.82) converges at low Z for any nonzero value of the parameter Y, while in the conventional theory the corresponding integral diverges at low Z. The physical reason is that the generalized theory takes into account the average splitting of the Stark sublevels caused by the z-component of the field of perturbing electrons (in addition to the allowance for the Stark splitting by the quasistatic part **F** of the ion microfield in plasmas). This additional splitting facilitates the convergence of the integral over the impact parameter at small values of Z (i.e. at small values of ρ).

The nonadiabatic part of the electron impact operator also contains the so-called interference term (for the non-Lyman spectral lines), which is nonzero only for the nondiagonal matrix elements of this operator:

$$\ll \alpha\beta \ \Phi_x \ \alpha'\beta' \gg = 2\frac{g}{a_0^2} \ x_{\alpha\alpha'} \ x^*_{\beta\beta'} \int_\infty^0 \frac{dZ_\alpha}{Z_\alpha} C_x(\chi_\alpha, Y_\alpha, Y_\beta, Z_\alpha, Z_\beta). \quad (7.89)$$

In equation (7.89), C_x is as follows:

$$C_x(\chi_\alpha, Y_\alpha, Y_\beta, Z_\alpha, Z_\beta) = \frac{3}{4}\int_\infty^\infty dx_1 \int_\infty^\infty dx_2 [w(x_1)w(x_2)]^3 exp[i(Z_\beta x_2 - Z_\alpha x_1)]$$

$$- \left\{ j_0\left(\varepsilon) + (2x_1x_2 1)\frac{j_1(\varepsilon)}{\varepsilon} + [(1x_1x_2)\sigma_1^2(x_1 + x_2)\sigma_1\sigma_2]\varepsilon^2 j_2(\varepsilon \right)\right\}, \quad (7.90)$$

$$\varepsilon \equiv \sqrt{\sigma_1^2 + \sigma_2^2}, \ \sigma_1 \equiv 2\frac{\chi_\alpha}{Z_\alpha} + (1 + x_2 w(x_2))\frac{Y_\beta}{Z_\beta} - (1 + x_1 w(x_1))\frac{Y_\alpha}{Z_\alpha},$$

$$\sigma_2 \equiv w(x_1)\frac{Y_\alpha}{Z_\alpha} - w(x_2)\frac{Y_\beta}{Z_\beta}, \ w(x) \equiv 1/\sqrt{1 + x^2}.$$

The quantity $C_\pm(\chi, Y, Z)$ from equation (7.84) can be broken down in two terms:

$$C_\pm(\chi, Y, Z) \equiv A_\pm(\chi, Y, Z) + iB_\pm(\chi, Y, Z) \quad (7.91)$$

In equation (7.91), the terms $A_\pm(\chi, Y, Z)$ and $B_\pm(\chi, Y, Z)$ are real quantities. The term $A_\pm(\chi, Y, Z)$ is called the width function, while the term $B_\pm(\chi, Y, Z)$ is called the shift function.

Figure 7.10 shows the Z-dependence of the width function $A-(Z)$ of the conventional theory (labeled in the figure as ST, 'standard theory') and of the width function $A-(Z,\kappa)$ of the generalized theory for several values of the parameter κ defined as

$$\kappa = q\,|\chi|\,Y, \quad (7.92)$$

where χ and Y are as given in equations (7.86) and (7.87), respectively. One can see that there is a very significant qualitative difference between the width function of the conventional theory and the width function of the generalized theory. In the generalized theory, it oscillates as Z approaches zero and vanishes at $Z = 0$. So, in the integral from equation (7.82), even after the width function is multiplied by $1/Z$, the integration converges at small Z due to the oscillations of the width function. However, this is not the case for the conventional theory, where the width function approaches unity (rather than zero) at small Z without oscillation. Therefore, after this width function is multiplied by $1/Z$, the integral in equation (7.82) diverges.

There are practical situations in which the nonadiabatic contribution to the electron-induced dynamical Stark width is much smaller than the adiabatic

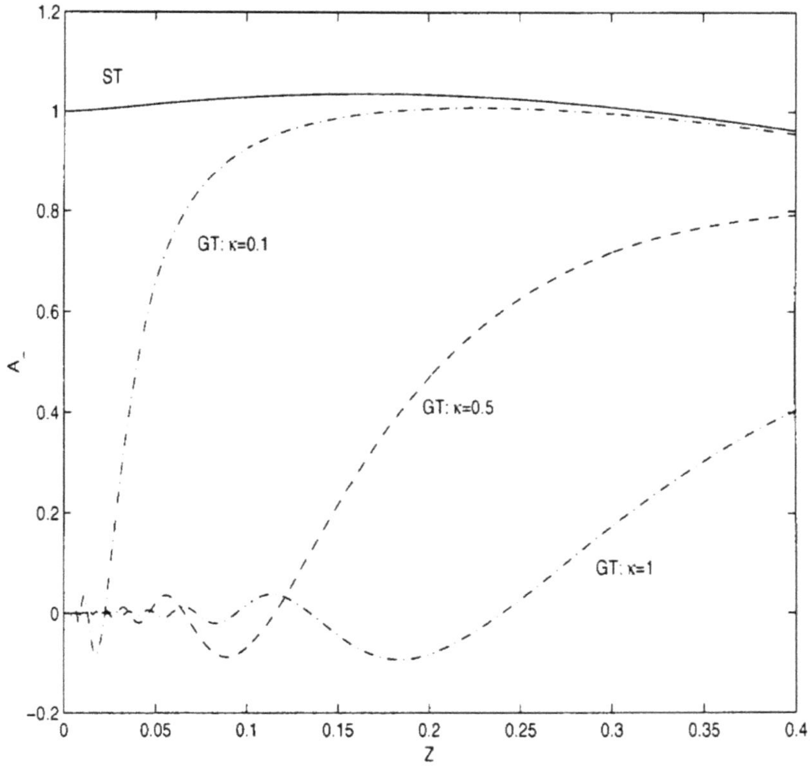

Figure 7.10. The Z-dependence of the width function $A-(Z)$ of the conventional theory (labeled in the figure as ST, 'standard theory') and of the width function $A-(Z,\kappa)$ of the generalized theory for several values of the parameter κ defined as $\kappa = q|\chi|\,Y$, where χ and Y are as given in equations (7.86) and (7.87), respectively. Reproduced from [18]. © Alpha Science International Limited.

contribution. For example, in magnetized plasmas, a sufficiently strong magnetic field **B** significantly suppresses the nonadiabatic contribution without affecting the adiabatic contribution, as shown in [32–35]. Therefore, in the following, we focus on the adiabatic contribution given by equations (7.79)–(7.81) and study its nonlinearities.

The parameter $u_{\alpha\beta}$ from equation (7.81) can be redefined in the form

$$u_{\alpha\beta} = m_e v^2/[3\hbar(n_\alpha q_\alpha - n_\beta q_\beta)\omega_{pe}], \tag{7.93}$$

where

$$\omega_{pe} = (4\pi e^2 N_e/m_e)^{1/2} \tag{7.94}$$

is the plasma electron frequency. From equation (7.93), it can be seen that the parameter $u_{\alpha\beta}$ can be interpreted as the scaled kinetic energy K_s of the perturbing electron. From equation (7.79), it can be seen that the adiabatic Stark width is controlled by the product $N_e\mathrm{Int}(u_{\alpha\beta}) = N_e\mathrm{Int}(K_s)$, where the function $\mathrm{Int}(u)$ is given by equation (7.81). At large u, the asymptotic expression for this function is

$$\mathrm{Int}(K_s) \approx 0.209 + \ln(K_s)/6. \tag{7.95}$$

Figure 7.11 presents the dependence of the factor Int on the scaled kinetic energy K_s of the perturbing electron (solid line). For comparison, the asymptotic result given by equation (7.95) is shown by the dashed line.

From figure 7.11, it can be seen that the dependence of the factor Int on the scaled kinetic energy K_s of the perturbing electron is nonlinear. It can also be seen that for $K_s > 1$, the exact and approximate expressions for the factor Int are very close to each other.

Here, we introduce the scaled adiabatic Stark width Γ_s, which is related to the (unscaled) adiabatic Stark width Γ as follows:

$$\Gamma_s = \Gamma_s m_e^{5/4} e^{1/2} / \left(3^{1/2} 6\pi^{1/4} \hbar^{3/2} X_{\alpha\beta}{}^{3/2}\right). \tag{7.96}$$

We also introduce the scaled electron temperature T_s, which is related to the (unscaled) electron temperature T_e as follows:

$$T_s = T_e m_e^{1/2} / \left(6\pi^{1/2} X_{\alpha\beta} \hbar e\right). \tag{7.97}$$

The scaled adiabatic Stark width from equation (7.79) can now be represented in the form

$$\Gamma_s = N_e\, \mathrm{Int}(T_s / N_e^{1/2}). \tag{7.98}$$

Figure 7.12 shows the dependence of the scaled adiabatic width Γ_s on the scaled electron temperature T_s and on the electron density N_e (cm^{-3}) for a relatively low range of electron densities.

Figure 7.13 presents the same plot as that of figure 7.12 but rotated by 90°, so that together with figure 7.12 it gives the most comprehensive view of the dependence of the scaled adiabatic width Γ_s on the scaled electron temperature T_s and on the electron density N_e at a relatively low range of N_e.

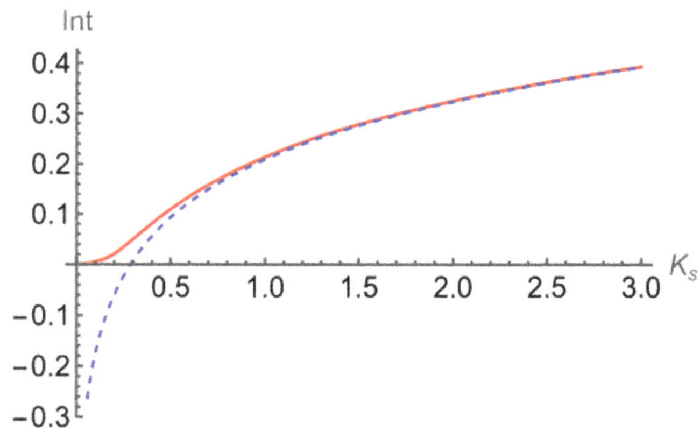

Figure 7.11. Dependence of the factor Int from equation (7.81) on the scaled kinetic energy K_s of the perturbing electron (solid line). For comparison, the asymptotic result given by equation (7.95) is shown by the dashed line.

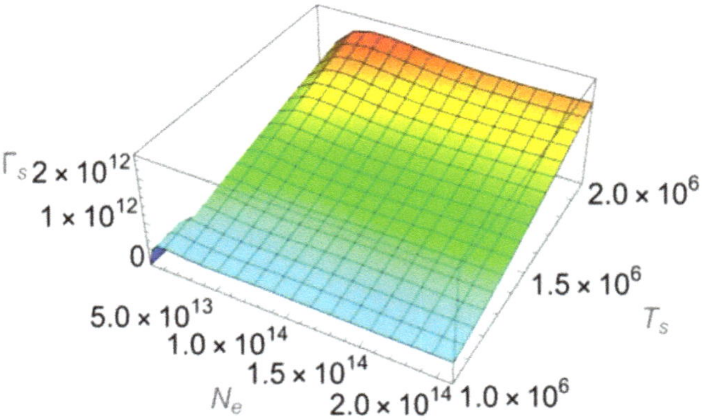

Figure 7.12. The dependence of the scaled adiabatic width Γ_s on the scaled electron temperature T_s and on the electron density N_e (cm^{-3}) for a relatively low range of electron densities.

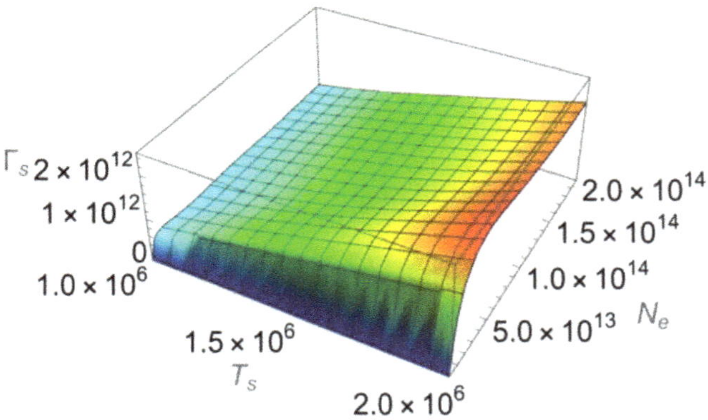

Figure 7.13. The same plot as in figure 7.12 but rotated by 90°, so that together with figure 7.12 it gives the most comprehensive view of the dependence of the scaled adiabatic width Γ_s on the scaled electron temperature T_s and on the electron density N_e (cm^{-3}) at a relatively low range of electron densities.

Figure 7.14 shows the dependence of the scaled adiabatic width Γ_s on the electron density N_e (cm^{-3}) at a relatively low range of electron densities for the scaled electron temperature $T_s = 10^6$.

Figure 7.15 shows the dependence of the scaled adiabatic width Γ_s on the scaled electron temperature T_s and on the electron density N_e (cm^{-3}) at a relatively high range of electron densities.

Figure 7.16 represents the same plot as that of figure 7.15 but rotated by 90°, so that together with figure 7.15 it gives the most comprehensive view of the dependence of the scaled adiabatic width Γ_s on the scaled electron temperature T_s and on the electron density N_e at a relatively high range of N_e.

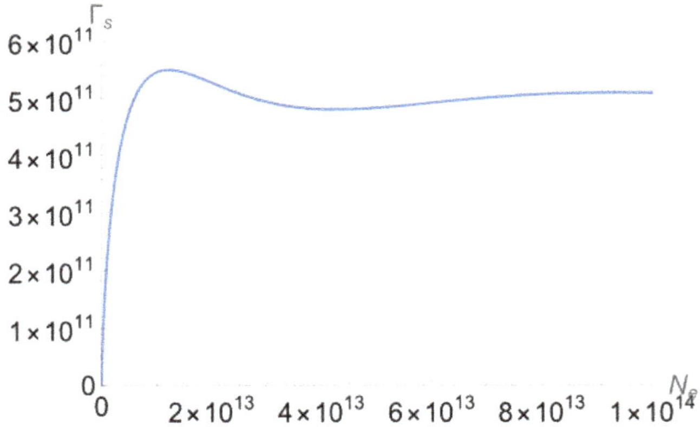

Figure 7.14. The dependence of the scaled adiabatic width Γ_s on the electron density N_e (cm^{-3}) at a relatively low range of electron densities for the scaled electron temperature $T_s = 10^6$.

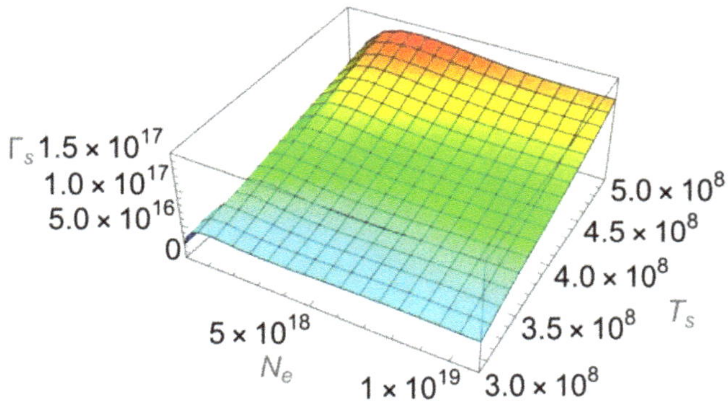

Figure 7.15. The dependence of the scaled adiabatic width Γ_s on the scaled electron temperature T_s and on the electron density N_e (cm^{-3}) at a relatively high range of electron densities.

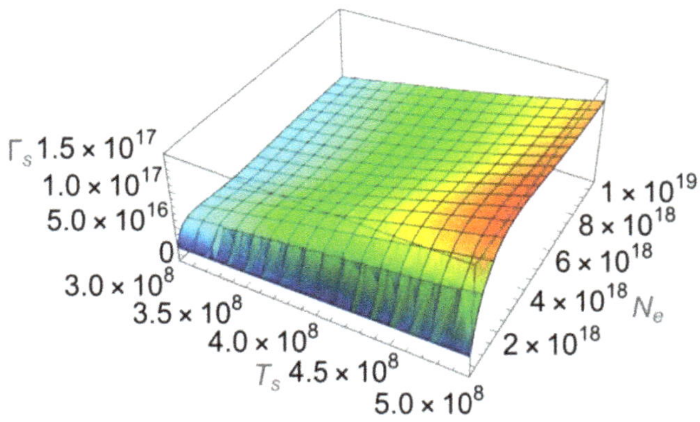

Figure 7.16. The same plot as that shown in figure 7.15 but rotated by 90°, so that together with figure 7.15 it gives the most comprehensive view of the dependence of the scaled adiabatic width Γ_s on the scaled electron temperature T_s and on the electron density N_e at a relatively high range of N_e.

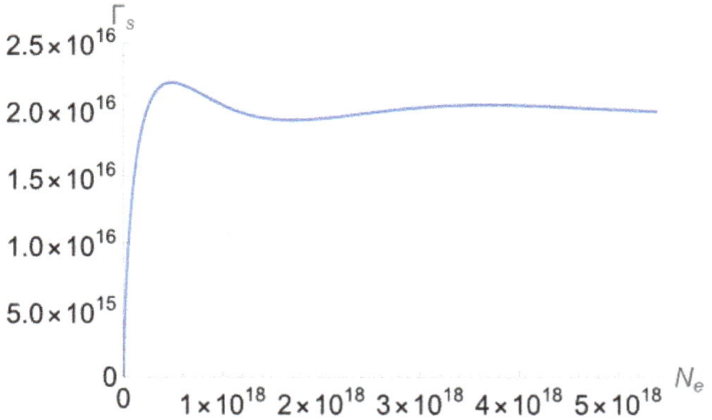

Figure 7.17. The dependence of the scaled adiabatic width Γ_s on the electron density N_e (cm^{-3}) at a relatively high range of electron densities for the scaled electron temperature $T_s = 2 \times 10^8$.

Figure 7.17 shows the dependence of the scaled adiabatic width Γ_s on the electron density N_e (cm^{-3}) at a relatively high range of electron densities for the scaled electron temperature $T_s = 2{\times}10^8$.

From figures 7.12–7.17, it can be seen that the dependence of the adiabatic Stark width (due to plasma electrons) on the electron density is significantly nonlinear.

7.5 Reduction of the Stark broadening due to the acceleration of electrons by the ion field in plasmas

The quasistatic part of the ion microfield in the vicinity of the radiator causes perturbing electrons to pass the radiator faster than when no allowance is made for the ion microfield. Because the dynamical Stark widths and shifts are inversely (approximately) proportional to the velocity of the perturbing electrons, this effect diminishes both the dynamical Stark shift and the dynamical Stark width. The lower the electron temperature and the higher the electron density, the more significant this decrease should become.

This effect was first presented analytically in [36] in frames of the binary approximation. Some analytical improvements were then made in [37]. This effect was later presented analytically using an approximate multiparticle description in [18] and [27].

This reduction of the Stark widths and shifts is a manifestation of the *direct* coupling between the ion and electron microfields. It is called the direct coupling to distinguish it from the coupling described in section 7.4. In the latter, *indirect* coupling, the radiator plays the role of an intermediary. The coupling parameter given by equation (7.87) depends on the parameters of the radiator (on its quantum numbers), on the parameter of the perturbing electrons (on their velocity), and on the parameter of the ion microfield (on the quasistatic field strength *F*).

Later, a group of authors performed simulations intended to reproduce this phenomenon [38]. Their simulations were limited to just one value of the electron temperature, just one value of the electron density, and just one spectral line

(Balmer-alpha). They claimed that they found an increase (rather than a decrease) in the Stark width caused by electrons.

The primary flaw of the fully-numerical approach of [38] is that its authors attempted to 'mimic' the binary version of this effect. Therefore, their results have no relevance to the analytical results obtained in the (approximate) multiparticle description [18, 27].

This controversy was ultimately resolved in [39], where we treated this phenomenon analytically within *statistical* mechanics—in contrast to our previous studies [27, 36, 37], where we employed a *dynamical* treatment of this phenomenon (specifically, where we first calculated the modification of the trajectory of a perturbing electron due to the ion field and then averaged the contribution of the perturbing electron to the Stark width over the ensemble of the perturbing electrons). In [39], our starting point was the modified distribution function of the electron velocities from Romanovsky–Ebeling [40]: the distribution modified by the ion microfield and calculated in [40] in the multiparticle description of the ion microfield. We then used this distribution to calculate the electron-caused Stark broadening within the conventional theory. The result demonstrated a decrease in the Stark broadening. So, the two different analytical formalisms yielded decreases in the Stark broadening and thus invalidated the simulation described in [38]. In the following, we give some details from [39].

The authors of paper [40] subdivided the instantaneous state of a plasma under consideration into a group of 'domains,' each having the size of about the Debye radius and each experiencing a different constant value of the quasistatic field **F** representing the ion microfield. The microfield was describe by the multiparticle formalism. The changes in the domain structure occurred at time intervals of the order of $1/\omega_{pe.}$ The authors of [40] employed the statistical approach to derive the corresponding (un-normalized) distribution of the plasma electron velocities:

$$f^{(u)}(v) = (2v/\pi)\int_0^\infty dt \sin(tv) \exp[-(v_F t)^{3/2} - (v_T t)^2/4]. \tag{7.99}$$

In equation (7.99), the superscript 'u' signifies 'un-normalized'; the quantity $v_T = (2T_e/m)^{1/2}$ is the electron's mean thermal velocity. The quantity v_F is the typical ion microfield scaled to the dimension of velocity:

$$v_F = (eE_H/m_e)[m_e/(4\pi e^2 N_i)]^{1/2}(T_i/T_e)^{1/4}. \tag{7.100}$$

In equation (7.100), the quantity E_H is the standard characteristic strength of the ion microfield:

$$E_H = 2\pi(4/15)^{2/3} e N_i^{2/3}. \tag{7.101}$$

In the limit of $v_E = 0$, the distribution from equation (7.99) simplifies to the Maxwell distribution, while in the limit of $v_T = 0$, the distribution from equation (7.99) simplifies to the Holtsmark distribution [41].

Obviously, the distribution function $f^{(u)}(v)$ from equation (7.99) is controlled by the ratio

$$v_F/v_T = (\pi/2)^{1/2}(4/15)^{2/3}eN_e^{1/6}/T_e. \quad (7.102)$$

Figure 7.18 shows the dependence of this ratio on the electron density N_e (in cm^{-3}) and on the electron temperature T_e (in eV).

Figure 7.19 presents the same plot as that of figure 7.18 but rotated by 90°, so that together with figure 7.18 it gives the most comprehensive view of the dependence of the ratio v_F/v_T on the electron density N_e (in cm^{-3}) and on the electron temperature T_e (in eV).

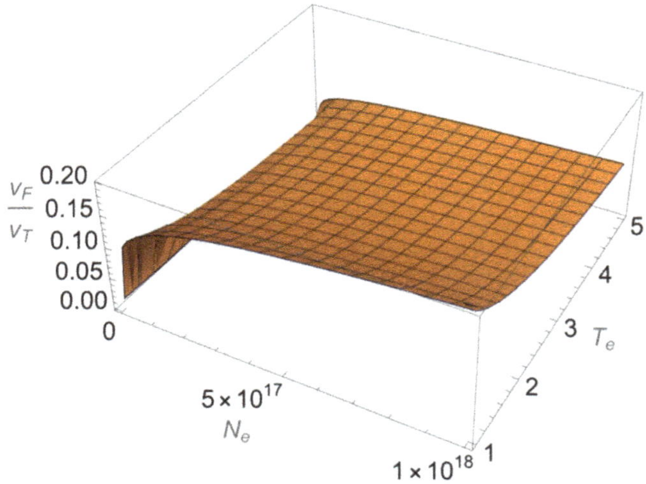

Figure 7.18. The dependence of the ratio v_F/v_T on the electron density N_e (in cm^{-3}) and on the electron temperature T_e (in eV).

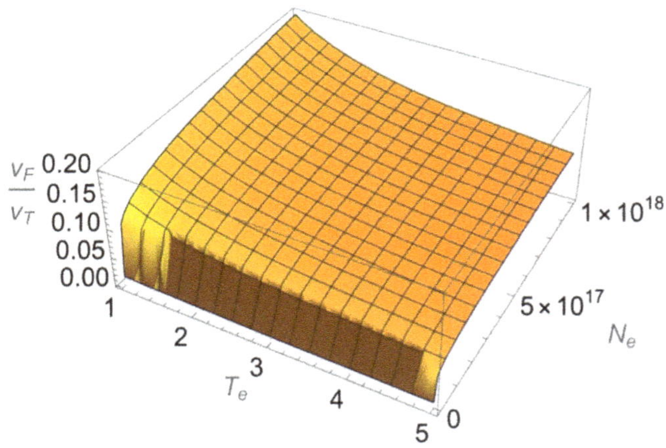

Figure 7.19. The same plot as in figure 7.18 but rotated by 90°, so that together with figure 7.18 it gives the most comprehensive view of the dependence of the ratio v_F/v_T on the electron density N_e (cm^{-3}) and on the electron temperature T_e (eV).

In paper [39], to obtain more explicit results, we analyzed the situation in which $v_F \ll v_T$, which is most important practically. We expanded $\exp[-(v_F\, t)^{3/2}]$ in equation (7.99):

$$f^{(u)}_{small}(v) = (2v/\pi)\int_0^\infty dt \sin(tv)\,[1 - (v_F t)^{3/2}]\exp[-(v_T t)^2/4]. \qquad (7.103)$$

In equation (7.103), the subscript 'small' signifies a relatively small ion field (i.e. a relatively small ion density). We analytically calculated the integral in equation (7.103) and then analytically normalized the resulting distribution to the form:

$$f_{small}(v) = [4\,v^2/(\pi v_T^{9/2})][\pi^{1/2} v_T^{3/2}\exp(-v^2/v_T^2) - 2^{5/2}\Gamma(9/4)\,v_F^{3/2} F(9/4,\,3/2;\,-v^2/v_T^2)]. \qquad (7.104)$$

In equation (7.104), $F(a,c;z)$ is the confluent hypergeometric function and $\Gamma(z)$ is the gamma function.

Before calculating the average over the electron velocities, the electron impact operator can be represented in the form

$$\Phi(v) = K/v^2, \qquad (7.105)$$

where the operator K is practically independent of the electron velocities (see, e.g. [41]):

$$K = -(\mathbf{r}_a\mathbf{r}_a - 2\mathbf{r}_a\mathbf{r}_b^* + \mathbf{r}_b^*\mathbf{r}_b^*)[4\pi e^4 N_e/(3\hbar^2)]\left[\ln(\rho_{max}/\rho_{min}) + 1/2\right]. \qquad (7.106)$$

We then averaged $\Phi(v)$ from equation (7.105) over the distribution of the electron velocities from equation (7.103),

$$\Phi_{small} = \int_0^\infty dv\, v f_{small}(v)\,\Phi(v), \qquad (7.107)$$

and obtained:

$$\Phi_{small} = [2K/(\pi^{1/2} v_T)][1 - 2^{3/2}\Gamma(5/4)\,v_F^{3/2}/(\pi^{1/2} v_T^{3/2})]. \qquad (7.108)$$

Equation (7.107) clearly proves that as the ratio v_E/v_T, which is the scaled dimensionless ion field, grows, the electron-induced Stark broadening *decreases*. Just this single result nullifies the statement in [38] that presumably the acceleration of the perturbing electrons by the ion field increases the Stark broadening by plasma electrons.

To study the nonlinearity of the electron impact operator Φ_{small} from equation (7.108), we scaled the electron density:

$$N_{e,\,scaled} = N_e \pi^3 (4/15)^4 e^6/m_e^3 = v_F^6. \qquad (7.109)$$

The scaled operator $\Phi_{s,small}$ can now be represented in the form:

$$\Phi_{s,\,small} = (1/v_T)[1 - 2^{3/2}\Gamma(5/4)\,N_{e,\,scaled}^{1/4}/(\pi^{1/2} v_T^{3/2})]. \qquad (7.110)$$

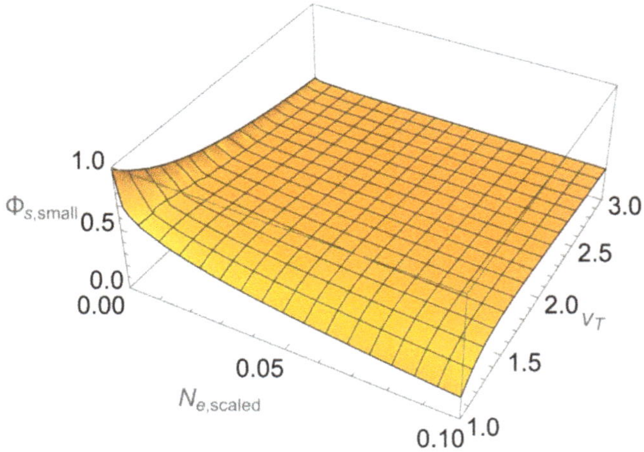

Figure 7.20. The dependence of the scaled electron impact operator $\Phi_{s,small}$ on the scaled electron density $N_{e,scaled}$ and on the mean thermal velocity of plasma electrons v_T.

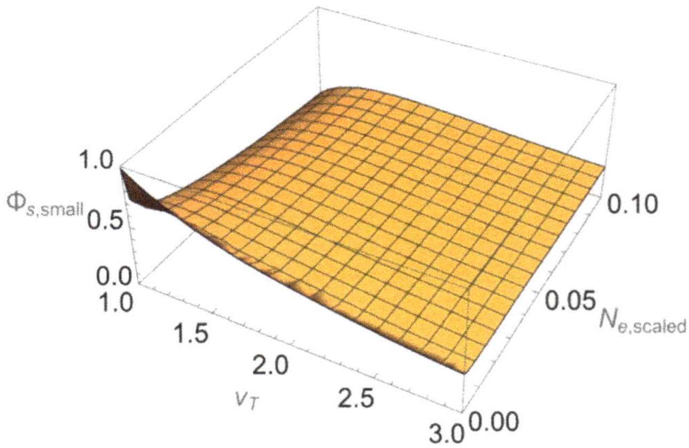

Figure 7.21. The same plot as that shown in figure 7.20 but rotated by 90°, so that together with figure 7.20 it gives the most comprehensive view of the dependence of the scaled electron impact operator $\Phi_{s,small}$ on the scaled electron density N_e and on the mean thermal velocity of plasma electrons v_T.

Figure 7.20 shows the dependence of the scaled electron impact operator $\Phi_{s,small}$ on the scaled electron density $N_{e,scaled}$ and on the mean thermal velocity of plasma electrons v_T.

Figure 7.21 presents the same plot as in figure 7.20 but rotated by 90°, so that together with figure 7.20 it gives the most comprehensive view of the dependence of the scaled electron impact operator $\Phi_{s,small}$ on the scaled electron density N_e and on the mean thermal velocity of plasma electrons v_T.

From figures 7.20 and 7.21, it can be seen that the electron broadening operator Φ_{small} depends on the electron density in a significantly nonlinear way.

We now analyze the opposite limit, where $v_F \gg v_T$. We expand $\exp[-(v_T\, t)^2/4]$ in equation (7.99):

$$f^{(u)}_{\text{large}}(v) = (2v/\pi)\int_0^\infty dt \sin(tv)\,[1 - (v_T t)^2/4]\exp[-(v_F t)^{3/2}]. \tag{7.111}$$

In equation (7.111), the subscript 'large' signifies a relatively large ion field. We perform the analytical integration in equation (7.111) and then analytically normalize the result to obtain the normalized velocity distribution function $f_{\text{large}}(v)$. We then employ $f_{\text{large}}(v)$ to obtain the average over velocities:

$$\Phi_{\text{large}} = \int_0^\infty dv\, v\, f_{\text{large}}(v)\, \Phi(v). \tag{7.112}$$

We perform the analytical integration in equation (7.112) and obtain the following:

$$\Phi_{\text{large}} = [9K\,\Gamma(11/3)/(20\pi v_F)]\left\{1 - 20\, v_T^2/[27\,\Gamma(11/3) v_F^2]\right\}. \tag{7.113}$$

To study the nonlinearity of the electron impact operator Φ_{large} from equation (7.113), we again scale the electron density according to equation (7.109) and represent the scaled operator $\Phi_{s,\text{large}}$ in the form:

$$\Phi_{s,\text{ large}} = (1/N_{e,\text{ scaled}}^{1/6})[1 - 20 v_T^2/(27\,\Gamma(11/3)\, N_{e,\text{ scaled}})^{1/3}]. \tag{7.114}$$

Figure 7.22 shows the dependence of the scaled electron impact operator $\Phi_{s,\text{large}}$ on the scaled electron density $N_{e,\text{scaled}}$ and on the mean thermal velocity of plasma electrons v_T.

Figure 7.23 presents the same plot as that of figure 7.22 but rotated by 90°, so that together with figure 7.22 it gives the most comprehensive view of the dependence of the scaled electron impact operator $\Phi_{s,\text{large}}$ on the scaled electron density N_e and on the mean thermal velocity of plasma electrons v_T.

From figures 7.22 and 7.23, it can be seen that the electron broadening operator Φ_{large} depends on the electron density in a significantly nonlinear way.

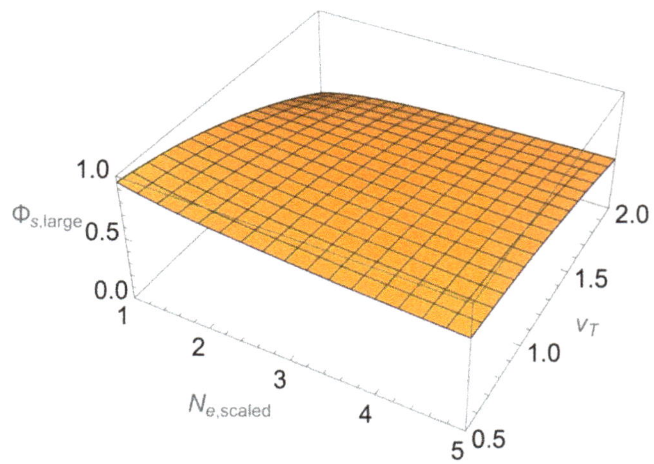

Figure 7.22. The dependence of the scaled electron impact operator $\Phi_{s,\text{large}}$ on the scaled electron density $N_{e,\text{scaled}}$ and on the mean thermal velocity of plasma electrons v_T.

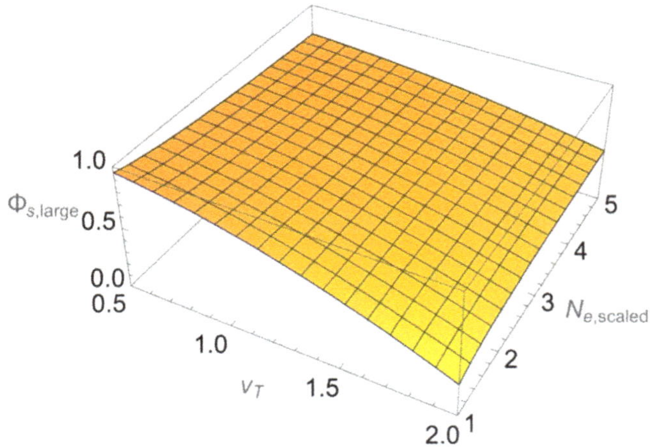

Figure 7.23. The same plot as that of figure 7.22 but rotated by 90°, so that together with figure 7.22 it gives the most comprehensive view of the dependence of the scaled electron impact operator $\Phi_{s,large}$ on the scaled electron density N_e and on the mean thermal velocity of plasma electrons v_T.

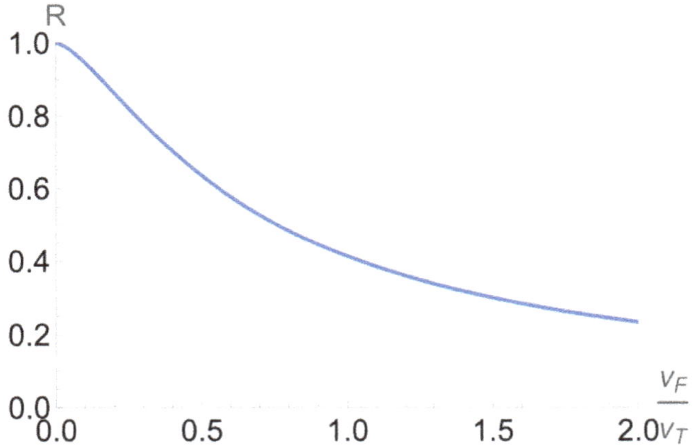

Figure 7.24. The reducing factor R from equation (7.115) versus the ratio v_F/v_T.

Finally, we define the factor R by which the electron-induced Stark broadening is reduced due to the acceleration of the perturbing electrons by the ion field:

$$R = \Phi(v_F/v_T)/\Phi(0). \tag{7.115}$$

Figure 7.24 shows this reducing factor R versus the ratio v_F/v_T.

From figure 7.24, one can see that the formalism of the statistical mechanics used in our paper [39] proves that the acceleration of the perturbing electron by the ion microfield *reduces* the Stark broadening. This is unequivocal proof that the simulations by the authors of [38], which reported the opposite effect, are incorrect.

We also note that the authors of paper [38] admitted the inability of their simulations to reproduce the analytical results from [42] dealing with the breakdown of the concept of the line space, while the authors of paper [38] did not find any flaws in the analytical calculations of [42]. In addition, it should be mentioned that another program developed by one of the authors of paper [38] (by Alexiou), which was similar to the code used in [38], produced a discrepancy of about 30% relative to the benchmark experiment [43]. All of the above create very significant doubts concerning the reliability of these programs.

Bibliography and additional reading

[1] Baranger M 1958 *Phys. Rev.* **111** 481
[2] Kolb A and Griem H R 1958 *Phys. Rev.* **111** 514
[3] Smith E, Cooper J and Vidal C 1969 *Phys. Rev.* **185** 140
[4] Vidal C, Cooper J and Smith E 1970 *J. Quant. Spectrosc. Radiat. Transf.* **10** 1011 1971**11** 263
[5] Smith E and Hooper C 1967 *Phys. Rev.* **157** 126
[6] Smith E 1968 *Phys. Rev.* **166** 102
[7] Lisitsa V S and Sholin G V 1972 *Sov. Phys. JETP* **34** 484 http://jetp.ras.ru/cgi-bin/e/index/e/34/3/p484?a=list
[8] Derevianko A and Oks E 1996 *Physics of Strongly Coupled Plasmas* ed W D Kraeft and M Schlanges (Singapore: World Scientific) 286 pp
[9] Oks E and Spectrosc J Q 2015 *Radiat. Transfer* **152** 74
[10] Griem H R 1974 *Spectral Line Broadening by Plasmas* (Cambridge, MA: Academic) https://shop.elsevier.com/books/spectral-line-broadening-by-plasmas/koch/978-0-12-302850-1
[11] Landau L D and Lifshitz E M 1965 *Quantum Mechanics* (Oxford: Pergamon)
[12] Cross R J Jr. 1967 *J. Chem. Phys.* **46** 609
[13] Fox K 1968 *J. Phys.* A **1** 124
[14] Chandrasekaran T and Wilkerson T W 1969 *Phys. Rev.* **181** 329
[15] Lebedev V S and Beigman I L 1998 *Physics of Highly Excited Atoms and Ions* (Berlin: Springer)
[16] Oks E 2020 *Analytical Advances in Quantum and Celestial Mechanics: Separating Rapid and Slow Subsystems* (Bristol: IOP Publishing)
[17] Griem H R and Shen K Y 1961 *Phys. Rev.* **122** 1490
[18] Oks E 2006 *Stark Broadening of Hydrogen and Hydrogenlike Spectral Lines in Plasmas: The Physical Insight* (Oxford: Alpha Science International)
[19] Oks E 2017 *Diagnostics of Laboratory and Astrophysical Plasmas Using Spectral Lines of One-, Two-, and Three-Electron Systems* (Hackensack, NJ: World Scientific)
[20] Sanders P and Oks E 2018 *J. Phys. Commun.* **2** 035033
[21] Galitski V, Karnakov B, Kogan V and Galitski V Jr. 2013 *Exploring Quantum Mechanics* (Oxford: Oxford University Press) problem 8.55
[22] Kotkin G L and Serbo V G 1971 *Collection of Problems in Classical Mechanics* (Oxford: Pergamon) problem 2.3
[23] Baryshenkov F F and Lisitsa V S 1981 *Sov. Phys. JETP* **53** 471 http://jetp.ras.ru/cgi-bin/e/index/e/53/3/p471?a=list
[24] Bureyeva L A and Lisitsa V S 2000 *A Perturbed Atom: Astrophysics and Space Physics Reviews* (Boca Raton, FL: CRC Press)
[25] Ispolatov Y and Oks E 1994 *J. Quant. Specrosc. Radiat. Transfer* **51** 129
[26] Oks E, Derevianko A and Ispolatov Y 1995 *J. Quant. Specrosc. Radiat. Transfer* **54** 307
[27] Oks E 2006 *AIP Conf. Proc.* **874** 19–34

[28] Oks E 1995 *Plasma Spectroscopy: The Influence of Microwave and Laser Fields* (Berlin: Springer)
[29] Iglesias C A, Lebowitz J and McGowan D 1983 *Phys. Rev.* A **32** 1667
[30] Lisitsa V S 1977 *Sov. Phys. Usp.* **122** 603
[31] Sholin G V, Demura A V and Lisitsa V S 1973 *Sov. Phys. JETP* **37** 1057 http://jetp.ras.ru/cgi-bin/e/index/e/37/6/p1057?a=list
[32] Derevianko A and Oks E 1994 *Phys. Rev. Lett.* **73** 2079
[33] Oks E 2010 *Int. Rev. Atom. Mol. Phys.* **1** 169
[34] Oks E 2012 Atomic processes in basic and applied physics *Springer Series on Atomic, Optical and Plasma Physics* vol 68 ed V Shevelko and H Tawara (New York: Springer) 393–431
[35] Oks E 2017 *J. Phys.: Conf. Ser.* **810** 012006
[36] Oks E 2000 *J. Quant. Spectr. Radiat. Transfer* **65** 405
[37] Oks E 2002 *J. Phys. B: At. Mol. Opt. Phys.* **35** 2251
[38] Stambulchik E, Fisher D V, Maron Y, Griem H R and Alexiou S 2007 *High Energy Density Phys.* **3** 272–7
[39] Oks E 2010 *Int. J. Spectrosc.* **2010** 347080
[40] Romanovsky M Y and Ebeling W 2003 *Phys. Lett.* A **317** 150
[41] Holtsmark J 1919 *Ann. Phys.* **58** 577
[42] Lee R W and Oks E 1998 *Phys. Rev. E, Rapid Communications* **58** 2441
[43] Büscher S, Wrubel T, Ferri S and Kunze H-J 2002 *J. Phys. B: At. Mol. Opt. Phys.* **35** 2889
[44] Lisitsa V S 1977 *Sov. Phys. Usp.* **122** 449

Chapter 8

Nonlinear effects in the ion-dynamical Stark broadening of hydrogenic spectral lines in plasmas

Whether the effect of the ion microfield on the Stark broadening of hydrogen spectral lines is mostly dynamical or quasistatic is controlled by the number of perturbing ions N_{wi} in the sphere of the Weisskopf radius r_{wi}:

$$N_{wi} = (4\pi N_i/3)r_{wi}^3, \tag{8.1}$$

where

$$r_{wi} = n^2\hbar/(m_e v_{Ti}). \tag{8.2}$$

In equations (8.1) and (8.2), N_i is the ion density, n is the principal quantum number of the upper energy level (from which the spectral line originates), m_e is the electron mass, and v_{Ti} is the mean thermal velocity of plasma ions:

$$v_{Ti} = (2T_i/m_i). \tag{8.3}$$

In equation (8.3), T_i is the ion temperature and m_i is the ion mass. In the following, we consider the plasma ions to be protons or deuterons, so that due to the quasi-neutrality of plasmas, one has

$$N_i = N_e, \tag{8.4}$$

where N_e is the electron density. We also consider the situation in which the electron temperature T_e is equal to T_i, so that we let

$$T = T_e = T_i. \tag{8.5}$$

When $N_{wi} \gg 1$, the effect of the ion microfield is mostly quasistatic. When $N_{wi} \ll 1$, the effect of the ion microfield is mostly dynamical.

doi:10.1088/978-0-7503-5552-0ch8

In this chapter, we are interested in the dynamical effect of the ion microfield, that is, when

$$N_{wi} \ll 1. \tag{8.6}$$

The condition (8.6) can be reformulated as follows. The plasma temperature T should exceed the critical value T_{cr}, where

$$T_{cr} = [4\pi N_e/(3N_{wi})]^{2/3} n^4 \hbar^2 m_i/m_e. \tag{8.7}$$

Figure 8.1 shows the dependence of the critical temperature T_{cr} on the electron density N_e (in cm^{-3}) and on the principal quantum number n for the number of ions in the ion Weisskopf sphere $N_{wi} = 1$.

Figure 8.2 displays the same plot as that of figure 8.1 but from the alternative viewpoint, so that together with figure 8.1 it shows the most comprehensive view of the dependence of the critical temperature T_{cr} on the electron density N_e (in cm^{-3}) and on the principal quantum number n for the number of ions in the ion Weisskopf sphere $N_{wi} = 1$.

Figure 8.3 shows the dependence of the critical temperature T_{cr} on the electron density N_e (in cm^{-3}) and on the number of ions in the ion Weisskopf sphere N_{wi} for the principal quantum number $n = 2$.

Figure 8.4 displays the same plot as that of figure 8.3 but from the alternative viewpoint, so that together with figure 8.1 it shows the most comprehensive view of the dependence of the critical temperature T_{cr} on the electron density N_e (in cm^{-3}) and on the number of ions in the ion Weisskopf sphere N_{wi} for the principal quantum number $n = 2$.

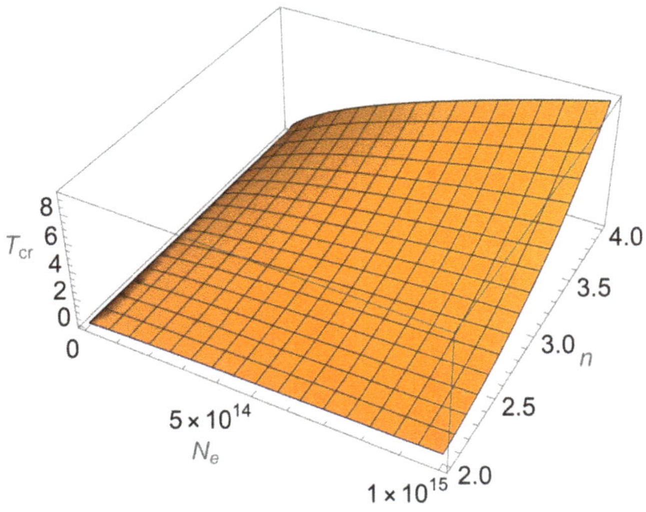

Figure 8.1. The dependence of the critical temperature T_{cr} on the electron density N_e (in cm^{-3}) and on the principal quantum number n for the number of ions in the ion Weisskopf sphere $N_{wi} = 1$.

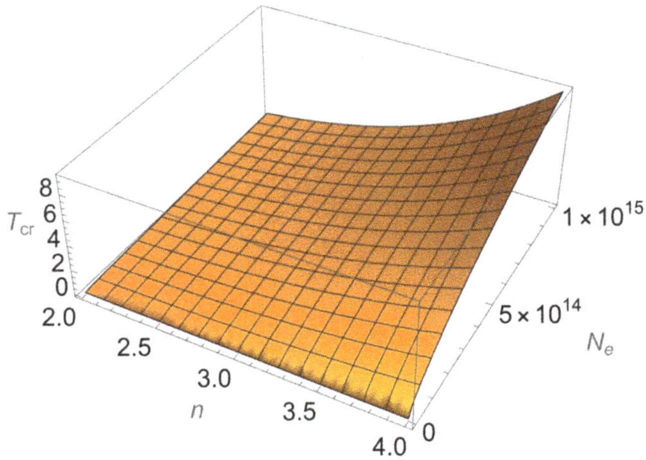

Figure 8.2. The same plot as that of figure 8.1 but from the alternative viewpoint, so that together with figure 8.1 it shows the most comprehensive view of the dependence of the critical temperature T_{cr} on the electron density N_e (in units of cm^{-3}) and on the principal quantum number n for the number of ions in the ion Weisskopf sphere $N_{wi} = 1$.

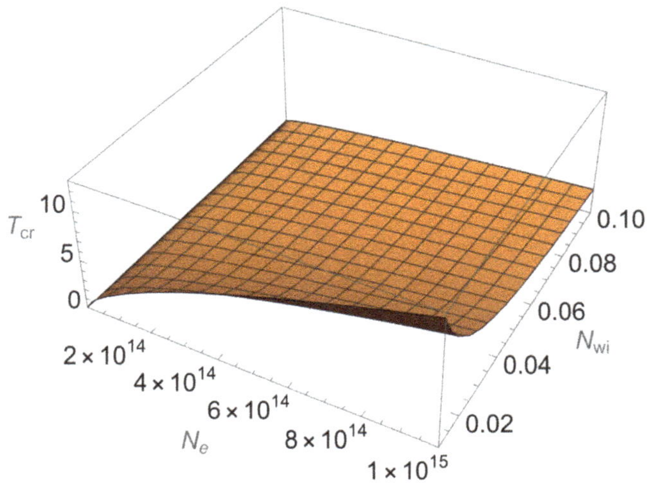

Figure 8.3. The dependence of the critical temperature T_{cr} on the electron density N_e (in cm^{-3}) and on the number of ions in the ion Weisskopf sphere N_{wi} for the principal quantum number $n = 2$.

Figure 8.5 shows the dependence of the critical temperature T_{cr} on the number of ions in the ion Weisskopf sphere N_{wi} and on the principal quantum number n for the electron density $N_e = 10^{14}$ cm^{-3}.

Figure 8.6 displays the same plot as that of figure 8.5 but from the alternative viewpoint, so that together with figure 8.5 it shows the most comprehensive view of the dependence of the critical temperature T_{cr} on the number of ions in the ion Weisskopf sphere N_{wi} and on the principal quantum number n for the electron density $N_e = 10^{14}$ cm^{-3}.

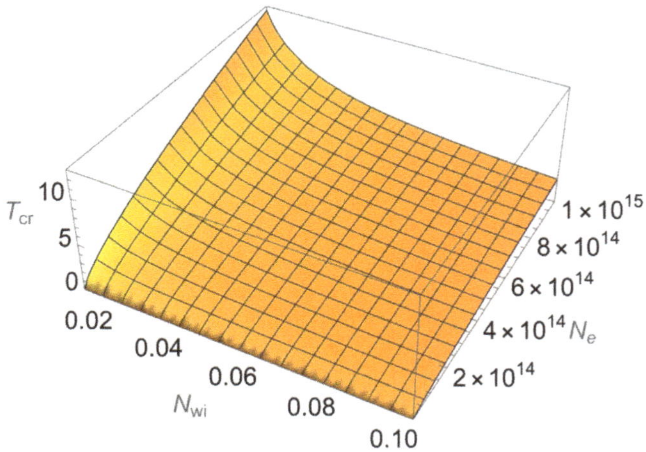

Figure 8.4. The same plot as that of figure 8.3 but from the alternative viewpoint, so that together with figure 8.1 it shows the most comprehensive view of the dependence of the critical temperature T_{cr} on the electron density N_e (in units of cm^{-3}) and on the number of ions in the ion Weisskopf sphere N_{wi} for the principal quantum number $n = 2$.

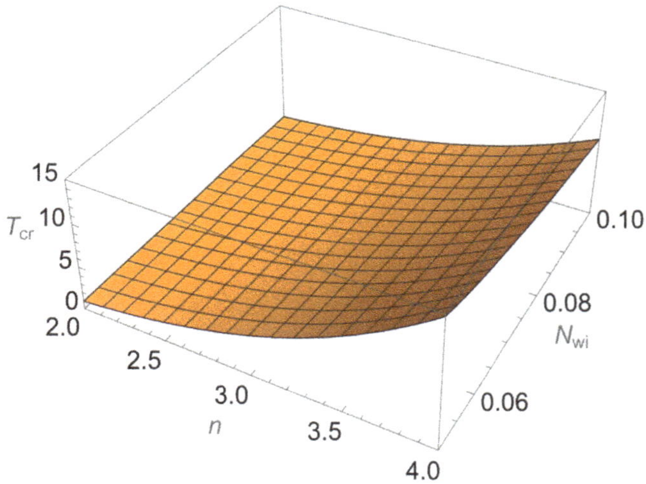

Figure 8.5. The dependence of the critical temperature T_{cr} on the number of ions in the ion Weisskopf sphere N_{wi} and on the principal quantum number n for the electron density $N_e = 10^{14}$ cm^{-3}.

We now focus on dynamical Stark broadening by the ion microfield not only in the conventional case, where $N_{wi} \ll 1$ (so that the ion microfield contributes mostly dynamically) but in the general case, including in particular $N_{wi} \sim 1$ and $N_{wi} \gg 1$. According to the generalized theory of Stark broadening, in the latter case, a minor part of the ion microfield still makes a dynamical contribution to Stark broadening. This contribution exhibits a counterintuitive dependence on plasma parameters, as presented below following [1].

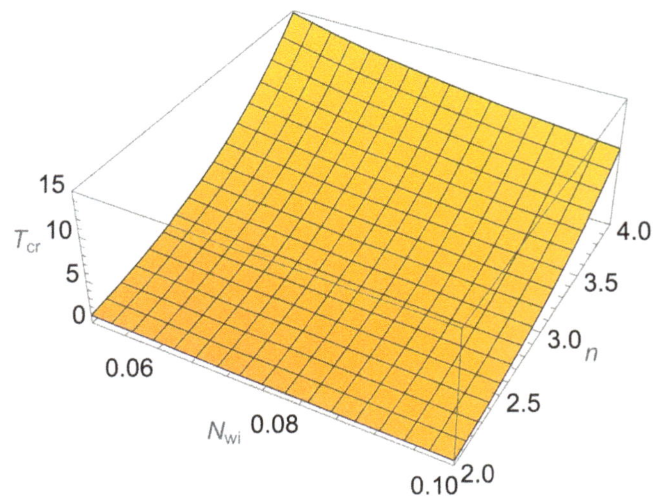

Figure 8.6. The same plot as that of figure 8.5 but from the alternative viewpoint, so that together with figure 8.5 it shows the most comprehensive view of the dependence of the critical temperature T_{cr} on the number of ions in the ion Weisskopf sphere N_{wi} and on the principal quantum number n for the electron density $N_e = 10^{14}$ cm^{-3}.

In the generalized theory of Stark broadening [2–4], the ion Weisskopf radius is not estimated by order of magnitude (as in equation (8.2)) but naturally arises from rigorous analytical derivation in the form

$$R_{wi} = 3X_{ab}\hbar/(m_e v_{Ti}), \tag{8.8}$$

where

$$X_{\alpha\beta} = n_\alpha(n_1 - n_2)_\alpha - n_\beta(n_1 - n_2)_\beta. \tag{8.9}$$

In equation (8.9), n, n_1, and n_2 are the principal and parabolic quantum numbers; the subscripts α and β denote the upper and lower Stark sublevels, respectively. As the right-hand side of equation (8.8) differs from the right-hand side of equation (8.2), we changed the notation on the left-hand side from r_{wi} to R_{wi}. Typically, the number of ions in the sphere of the ion Weisskopf radius

$$N_{wi} = (4\pi N_i/3)R_{wi}^3 \tag{8.10}$$

is much greater than unity, so that the greater part of the ion microfield affects the spectral lineshape quasistatically.

In this situation, that is, where $N_{wi} \gg 1$, we analyzed the dynamical contribution to the Stark broadening (due to a minor part of the ion microfield) using the following properties of the generalized theory. First, the adiabatic part of the dynamical contribution predominates over the nonadiabatic part (for $N_{wi} \gg 1$), so that the latter was disregarded. Second, the adiabatic contribution was exactly derived in the generalized theory (without using the perturbation theory). The analytical outcome presented in [1] was the ion dynamical contribution to the Stark broadening derived without engaging the binary approximation or the impact approximation or by assuming the collisions to be complete. Some details are offered below.

For a given argument τ of the correlation function $C(\tau)$, the ion dynamical contribution is due to collisions such that the instant t_0 of the closest approach to the radiator satisfies the inequality

$$-\tau/2 < t_0 < \tau/2. \tag{8.11}$$

We do not presume that $\tau \gg \rho/v$, that is, we do not engage the assumption of completed collisions. We keep the ratio $v\tau/\rho$ arbitrary, so that we take into account the incomplete collisions.

The ion dynamical part of the correlation function can be represented in the form

$$C(\tau) \approx \exp[-N_i V(\tau)], \quad V(\tau) \approx 2\pi v \int_0^\infty \mathrm{d}\rho\, \rho \langle\{1 - \exp[\mathrm{i}I(v, \rho, \tau)]\}\rangle \int_{-\tau/2}^{\tau/2} \mathrm{d}t_0, \tag{8.12}$$

where

$$\begin{aligned} I(v, \rho, \tau) &= (vR_{\mathrm{wi}}/2) \int_{-\tau/2}^{\tau/2} \mathrm{d}t E(v, \rho, t) = (\cos\theta R_{\mathrm{w}}/\rho) \int_0^{v\tau/(2\rho)} \mathrm{d}x (1 + x^2)^{-3/2} \\ &= G(v, \rho, \tau)\cos\theta. \end{aligned} \tag{8.13}$$

In equation (8.12), the symbol $\langle\cdots\rangle$ signifies the angular average. In equation (8.13), θ is the angle between the vectors ρ and $\mathbf{F}$ (where $\mathbf{F}$ is the quasistatic part of the ion microfield at the location of the radiator), and

$$G(v, \rho, \tau) = v\tau R_{\mathrm{wi}}/[2\rho(\rho^2 + v^2\tau^2/4)^{1/2}]. \tag{8.14}$$

After the angular average is obtained, the symbol $< \cdots >$ in equation (8.12) takes the form

$$\langle\cdots\rangle = 1 - [\sin G(v, \rho, \tau)]/G(v, \rho, \tau). \tag{8.15}$$

We then expressed ρ via G by utilizing equation (8.15) and changed the variable of the integration in equation (8.12) to be G instead of ρ. As a result, we found the final expression for the ion dynamical part of the correlation function to be the following:

$$\begin{aligned} C(\tau) &= \exp[-6N_{\mathrm{wi}} f(\Omega_{\mathrm{w}}\tau/4)], \\ f(x) &= \int_0^\infty \mathrm{d}y[1 - (\sin y)/y]/[y^2(y^2 + 1/x^2)^{1/2}]. \end{aligned} \tag{8.16}$$

In equation (8.16),

$$\Omega_{\mathrm{w}} = v/R_{\mathrm{wi}} \tag{8.17}$$

is the Weisskopf frequency.

Expression (8.16) for the correlation function is valid for any value of the product $\Omega_{\mathrm{w}}\tau$. Figure 8.7 shows the dependence of the correlation function C on $\Omega_{\mathrm{w}}\tau$ and on the number of ions N_{wi} in the sphere of the ion Weisskopf radius.

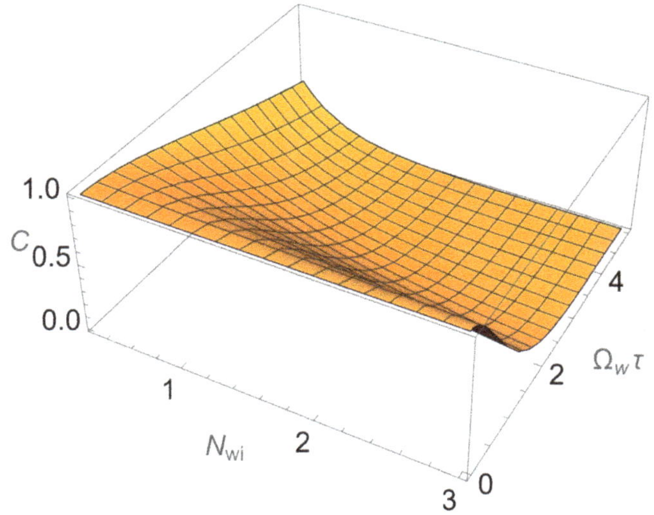

Figure 8.7. The dependence of the correlation function C on $\Omega_w\tau$ and on the number of ions N_{wi} in the sphere of the ion Weisskopf radius.

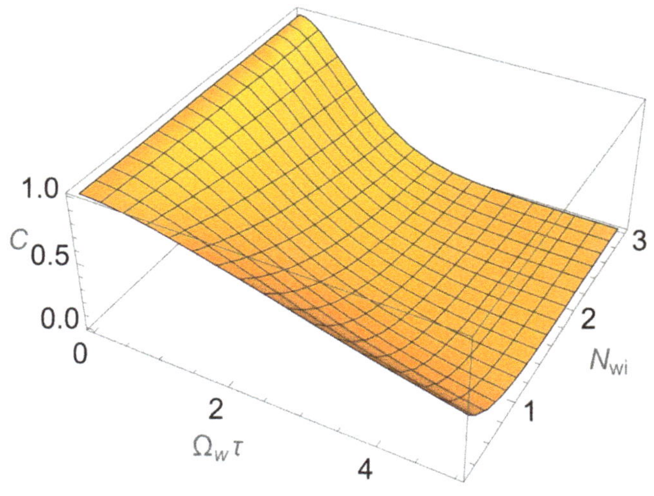

Figure 8.8. The same plot as that of figure 8.7 but from the alternative viewpoint, so that together with figure 8.7 it shows in the most comprehensive way the dependence of the correlation function C on $\Omega_w\tau$ and on the number of ions N_{wi} in the sphere of the ion Weisskopf radius.

Figure 8.8 displays the same plot as in figure 8.7 but from the alternative viewpoint, so that together with figure 8.7 it shows in the most comprehensive way the dependence of the correlation function C on $\Omega_w\tau$ and on the number of ions N_{wi} in the sphere of the ion Weisskopf radius.

From figures 8.7 and 8.8, it can be seen that for $N_{wi} \ll 1$, the correlation function falls off with the growth of τ significantly slower than for $N_{wi} > 1$.

In the limiting case, in which

$$\Omega_w \tau \gg 1, \tag{8.18}$$

the correlation function simplifies to

$$C(\tau) = \exp[-N_{wi}(\Omega_w \tau/4)\ln(\Omega_w \tau/4)]. \tag{8.19}$$

This correlation function leads to a *quasi-Lorentzian* lineshape.

In the opposite limiting case, in which $\Omega_w \tau \ll 1$ (corresponding to 'totally' incomplete collisions), the correlation function becomes

$$C(\tau) = \exp[-(3\pi/32)N_{wi}(\Omega_w \tau)^2]. \tag{8.20}$$

In this limit, the correlation function yields a *Gaussian* lineshape. The result, namely that dynamical Stark broadening yields the Gaussian lineshape, is *counterintuitive*.

So, in the situation in which $N_{wi} \gg 1$, the *total* correlation function and the corresponding lineshape have three different regions, as follows. First, there is the quasistatic region: $\Omega_w \tau \ll 1/N_{wi}^{1/2}$. Second, there is the Gaussian (dynamic) region: $1/N_{wi}^{1/2} \ll \Omega_w \tau \ll 1$. Third, there is the quasi-Lorentzian (dynamic) region: $1 \ll \Omega_w \tau$.

Figure 8.9 shows the dependence of the scaled ion dynamical Stark width $\Delta\omega_{1/2}/\Omega_w$ on the scaled ion density N_i/N_0, where N_0 corresponds to the ion density, for which the number of ions in the sphere of the ion Weisskopf radius is equal to unity.

From figure 8.9, it can be seen that the ion dynamical Stark width depends on the electron density in a significantly nonlinear way. At $N_i \ll N_0$, the dependence is quasi-linear. However, as the ion density exceeds the value ~$0.5N_0$, the dependence becomes significantly nonlinear: at $N_i \gg N_0$, it becomes $(N_i/N_0)^{1/2}$.

Figure 8.10 shows the dependence of the scaled ion dynamical Stark width $\Delta\omega_{1/2}/\Omega_w$ on the scaled ion temperature T/T_0, where T_0 corresponds to the ion

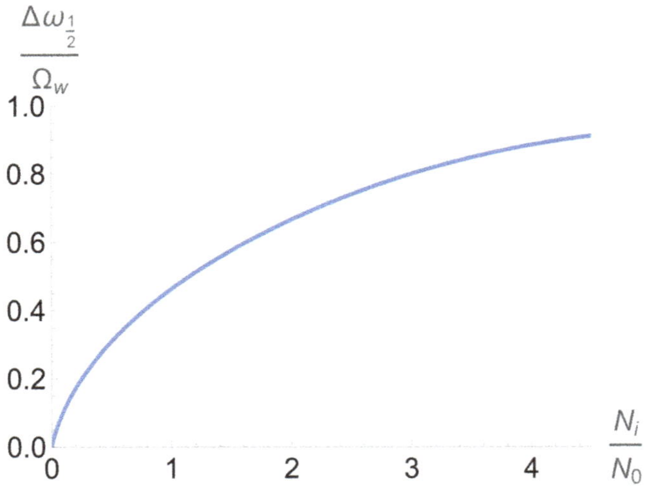

Figure 8.9. The dependence of the scaled ion dynamical Stark width $\Delta\omega_{1/2}/\Omega_w$ on the scaled ion density N_i/N_0, where N_0 corresponds to the ion density, for which the number of ions in the sphere of the ion Weisskopf radius is equal to unity.

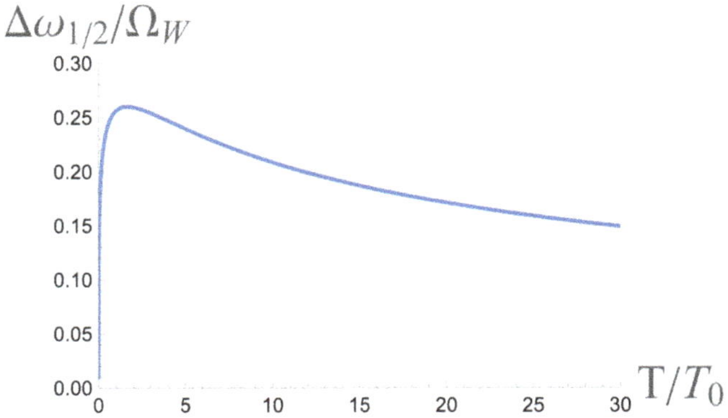

Figure 8.10. The dependence of the scaled ion dynamical Stark width $\Delta\omega_{1/2}/\Omega_w$ on the scaled ion temperature T/T_0, where T_0 corresponds to the ion temperature, for which the number of ions in the sphere of the ion Weisskopf radius is equal to unity.

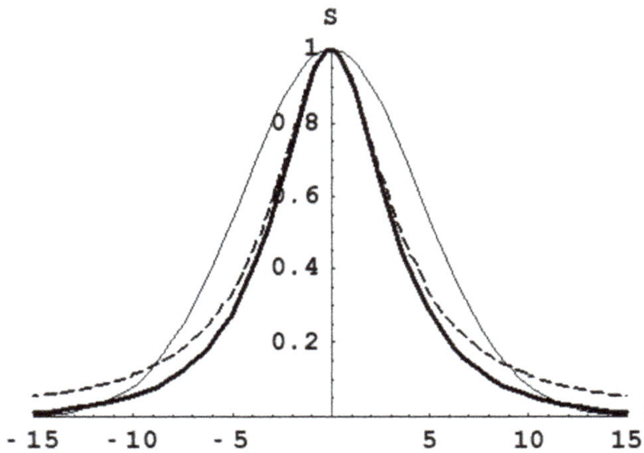

Figure 8.11. The typical profile of any *lateral* Stark component for dynamical broadening by the ion microfield. Our analytical result is shown by the bold solid line. The Lorentzian, matching the central part of our calculated profile, is shown by the dashed line. The Gaussian, matching our calculated profile in the wings, is shown by the thin solid line.

temperature, for which the number of ions in the sphere of the ion Weisskopf radius is equal to unity.

Figure 8.10 illustrates the following. At $T \gg T_0$, the scaled ion dynamical Stark width is $(T_0/T)^{1/2}$. As the temperature T decreases, the scaled width reaches its maximum at $T/T_0 = 1.64$ and then decreases. At $T \ll T_0$, the scaled width becomes $(T/T_0)^4$.

Figure 8.11 shows a typical profile of any *lateral* Stark component for dynamical broadening by the ion microfield. Our analytical result is shown by the bold solid line. The Lorentzian, matching the central part of our calculated profile, is shown by the dashed line. The Gaussian, matching our calculated profile in the wings, is shown by the thin solid line.

In equations (8.10)–(8.12) and in the subsequent results, it was assumed that $X_{\alpha\beta}$ (given by equation (8.9)) is nonzero, meaning that we considered *lateral* Stark components of hydrogen spectral lines. The situation in which the hydrogen line is dominated by the central (unshifted) Stark component, which is the case for the Ly-alpha line, was studied by Stambulchik and Demura [5]. Using simulations, they demonstrated that the Stark width for the Ly-alpha line is proportional to N_p at relatively small perturber (e.g. ion) densities N_p (the impact limit), but at large N_p it is proportional to $N_p^{1/3}$ (while for the quasistatic case, the Stark width would be proportional to $N_p^{2/3}$). They correctly interpreted this result by the consideration that for the central Stark component of the Ly-alpha line (the component that has 2/3 of the total intensity of the Ly-alpha line), the so-called amplitude modulation (of the atomic oscillator) is the dominating broadening mechanism for high N_p. So, the Stark width of the Ly-alpha line at relatively large perturber densities N_p is proportional to the rotational frequency of the perturbers, the rotational frequency being $\sim v_p N_p^{1/3}$. In the same way, the authors of [5] demonstrated that for the Ly-alpha line, the Stark width is proportional to $1/T^{1/2}$ at relatively large temperatures T (the impact result), but at relatively small temperatures, it becomes proportional to $T^{1/2}$ (i.e. it is again proportional to the rotational frequency of the perturbers $\sim v_p N_p^{1/3}$).

Bibliography

[1] Flih S A, Oks E and Vitel Y 2003 *J. Phys. B: At. Mol. Opt. Phys.* **36** 283

[2] Ispolatov Y and Oks E 1994 *J. Quant. Specrosc. Radiat. Transfer* **51** 129

[3] Oks E, Derevianko A and Ispolatov Y 1995 *J. Quant. Specrosc. Radiat. Transfer* **54** 307

[4] Oks E 2006 *AIP Conf. Proc.* **874** 19

[5] Stambulchik E and Demura A V 2016 *J. Phys. B: At. Mol. Opt. Phys.* **49** 035701

Appendix A

The effects of various electric fields on the 5D, 5F, and 5G energy levels of helium

The simplest three-level scheme employed by Baranger and Mozer [1] to describe the satellites of dipole-forbidden spectral lines under a quasimonochromatic electric field (QEF) of frequency ω and amplitude E_0 was applied by Kunze and Griem [2] to observe the satellites of the 492.2 nm (singlet) and 447.1 nm (triplet) helium lines. For these lines, the allowed transition is 4D–2P and the forbidden transition is 4F–2P. In order for the QEF to sufficiently intermix the levels 4D and 4F, thus causing the satellites, the QEF amplitude required to practically observe the satellites of these two helium lines is relatively high: ~10 kV cm^{-1}.

For many practical purposes, such as the spectroscopic diagnostic of the QEF at the lower hybrid frequency in tokamaks, in which the QEF amplitude is expected to be ~(1–4) kV cm^{-1}, it is necessary to increase the sensitivity of the helium atoms to the QEF. The following dimensionless parameter α controls the relative intensities of the satellite, the total number of satellites that can be observed, and most importantly, the sensitivity of this method:

$$\alpha = 2dE_0/(\hbar\Delta). \tag{A.1}$$

Here, d is the dipole matrix element coupling the levels, which originates the allowed spectral line and the satellites of the forbidden spectral line and Δ is the separation (in the angular frequency scale) between these levels. Situations for which $\alpha \ll 1$ can be described by the perturbation theory (as used by Baranger and Mozer [1]). However, for situations in which α is greater than or of the order of unity, the perturbation theory becomes inapplicable.

In order to increase the sensitivity parameter α at a fixed QEF amplitude E_0, the only way is to proceed to spectral lines originating from levels of the principal quantum number $n > 4$. Indeed, α is proportional to the ratio d/Δ. As n increases, d also increases (proportionally to n^2), while the separation Δ between closely lying levels decreases.

doi:10.1088/978-0-7503-5552-0ch9

Selecting prospective spectral lines of helium involves finding an optimal balance between several competing, counter-acting factors, such as the *sensitivity* to the anticipated QEF, the required *spectral resolution*, and sufficient *absolute intensity*. Indeed, on the one hand, the higher the principal quantum number n of the upper level, the better the sensitivity with respect to the electric fields. On the other hand, the higher the value of n, the smaller the separation between the forbidden and allowed components (or between satellites and the allowed component) of the corresponding spectral lines, thus requiring a higher spectral resolution. Also, as n increases, the absolute intensity of the spectral lines decreases. It should be also emphasized that a spectral line of helium, sensitive enough to the QEF ~ (1–4) kV cm^{-1}, should be chosen such that the effect of the QEF is strong enough that it cannot be described by the perturbation theory.

The three-level scheme theoretically employed by Baranger and Mozer [1] (and later by Cooper and Ringler [3]) was relatively simple because the QEF was supposed to intermix only two levels: 4D and 4F. (We recall that there are no levels 4G, 4H. etc.) Since enhancing the sensitivity requires proceeding to helium spectral lines originating from levels of $n > 4$, this means that the QEF will couple more than two levels. For example, if one were to use helium spectral lines in the vicinity of the allowed transition 5D–2P, then there would be two forbidden transitions: 5F–2P and 5G–2P. In other words, in this case, more sophisticated analytical approaches—going beyond the perturbation theory—should be employed for the four-level scheme whose three upper levels are 5D, 5F, and 5G.

The need may also arise to use helium spectral lines in the vicinity of the allowed transition 6D–2P. There would then be three forbidden transitions: 6F–2P, 6G–2P, and 6H–2P. In other words, in this case, the more sophisticated analytical approaches (beyond the perturbation theory) should be implemented for the five-level scheme whose four upper levels are 6D, 6F, 6G, and 6H.

Specifically, the two following analytical theories presented in [4] are relevant for this purpose. Both theories consider satellites of dipole-forbidden spectral lines under a linearly-polarized electric field $E(t) = E_0 \cos(\omega t)$. Their regions of validity, while differing, actually complement each other. For exactness, we provide below details for helium spectral lines corresponding to the radiative transitions to the lower level 2P from the upper levels (or sublevels) nD, nF, … . The total number of upper sublevels is $n - 2$, where n is the principal quantum number of the upper sublevels. Atomic units are used in the subsequent formulas.

The first theory is called the adiabatic theory of satellites. Its region of validity can be expressed as follows:

$$(zE_0/\Delta)(\omega/\Delta) \ll 1. \tag{A.2}$$

Here, z is the average of the matrix elements of the z-coordinate (chosen along the oscillatory electric field) between sublevels nD and nF, nF, and nG, etc.; Δ is the separation between the highest and the lowest of the sublevels nD, nF, … . For example, the adiabatic theory of satellites allows situations in which the amplitude E_0 is relatively high (so that the standard perturbation theory fails) to be treated analytically as long as the frequency ω is relatively low.

The second theory is called the high-frequency or strong-field theory of satellites. As follows from its name, it is valid where either the amplitude E_0 or the frequency ω is sufficiently large. Its region of validity can be expressed as follows:

$$(zE_0/\Delta)^2 + (\omega/\Delta)^2 \gg 1. \tag{A.3}$$

Figure A.1 schematically illustrates the regions of validity of the adiabatic theory of satellites and the high-frequency or strong-field theory of satellites. The former theory is valid below the solid curve, while the latter theory is valid above the dashed curve.

We now specifically consider the situation in which the three upper energy levels are 5D, 5F, and 5G. In figure A.2, these levels are denoted by numbers from one to three for the brevity of the subsequent equations and the analytical results.

In the following, we consider the subset of the energy levels corresponding to the zero projection of the angular momentum in the direction of the electric field: $m = 0$. They are coupled by the matrix elements of the z-coordinate, where (in atomic units)

$$z_{12} = 11.34, \quad z_{13} = 15.21. \tag{A.4}$$

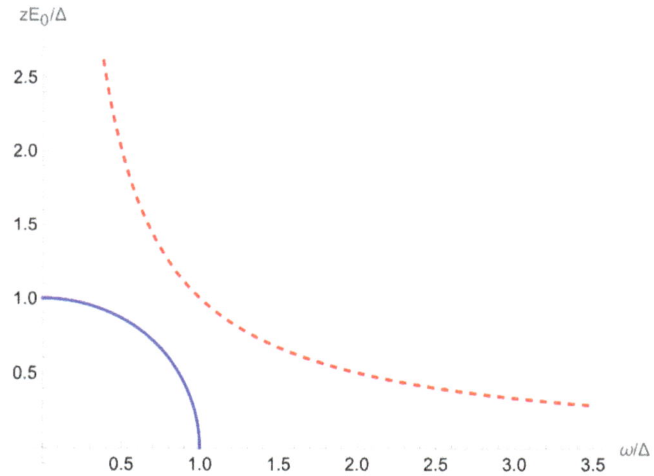

Figure A.1. Regions of validity of the adiabatic theory of satellites and the high-frequency or strong-field theory of satellites. The former theory is valid below the solid curve, while the latter theory is valid above the dashed curve.

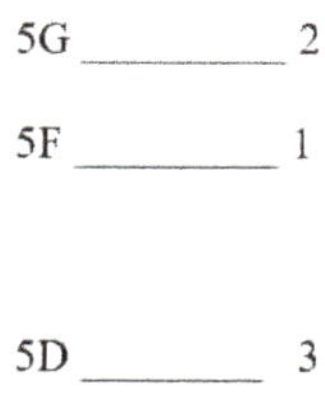

Figure A.2. The situation in which the three upper energy levels of helium in the four-level scheme are 5D, 5F, and 5G.

The energy differences in atomic units are the same as the frequency differences, which are as follows:

$$\omega_{21} = 2.237 \times 10^{-6}, \quad \omega_{31} = -1.294 \times 10^{-5}, \\ \omega_{23} = \omega_{21} - \omega_{31} = 1.518 \times 10^{-5}. \tag{A.5}$$

We introduce the following notations:

$$\alpha \equiv B \sin \omega t,\ B \equiv \xi E_0/\omega,\ \xi \equiv (z_{12}^2 + z_{13}^2)^{1/2}, \\ \tilde{z}_{12} \equiv z_{12}/\xi,\ \tilde{z}_{13} \equiv z_{13}/\xi, \\ V \equiv \omega_{21}\tilde{z}_{12}^2 + \omega_{31}\tilde{z}_{13}^2,\ W = \omega_{21}\tilde{z}_{13}^2 + \omega_{31}\tilde{z}_{12}^2. \tag{A.6}$$

The primary parameter controlling the quasienergies is the quantity B from equation (A.6). The practical formula for the parameter B is:

$$B = 24.28 E_0/\omega, \tag{A.7}$$

where the field amplitude E_0 is in kV cm^{-1} and the field frequency ω is in GHz.

Using the Fourier expansions

$$\sin \alpha = 2\sum_{k=1}^{\infty} J_{2k-1}(B)\sin(2k - 1)\omega t, \quad \cos \alpha = J_0(B) \\ + 2\sum_{k=1}^{\infty} J_{2k}(B)\sin 2k\omega t, \tag{A.8}$$

where $J_q(B)$ are the Bessel functions, the quasienergies μ_1, μ_2, and μ_3 can be obtained in the following form [4]:

$$\mu_1 = a_{11},\quad \mu_{2,3} = (a_{22} + a_{33})/2 \pm [(a_{22} - a_{33})^2/4 + a_{23}^2]^{1/2}. \tag{A.9}$$

In equation (A.9), the quantities a_{11}, a_{22}, a_{33}, and a_{23} are as follows:

$$a_{11} = [1 - J_0(2B)]V/2; \\ a_{23} = a_{32} = \tilde{z}_{12}\tilde{z}_{13}\{(\tilde{z}_{13}^2 - \tilde{z}_{12}^2)\omega_{23}[J_0(B) - 1] + [J_0(2B) - 1]V/2\}; \\ a_{pp} = \omega_{p1} + (-1)^p 2\tilde{z}_{12}^2\tilde{z}_{13}^2\omega_{23}[J_0(B) - 1] \\ + \tilde{z}_{1p}^2[J_0(2B) - 1]V/2,\ (p = 2,3). \tag{A.10}$$

ED denotes the scaled dimensionless energy density of the oscillatory field:

$$\text{ED} = B^2. \tag{A.11}$$

Figure A.3 presents the quasienergies of the energy levels from figure A.2 versus the scaled energy density ED of the oscillatory field.

From figure A.3, one can see that the dependence of the quasienergies on the scaled energy density of the oscillatory field is highly nonlinear.

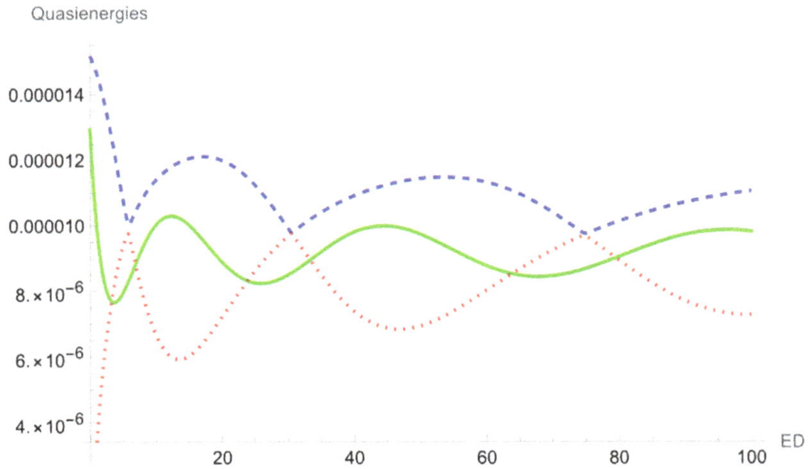

Figure A.3. Quasienergies of the energy levels from figure A.2 versus the scaled energy density ED of the oscillatory field.

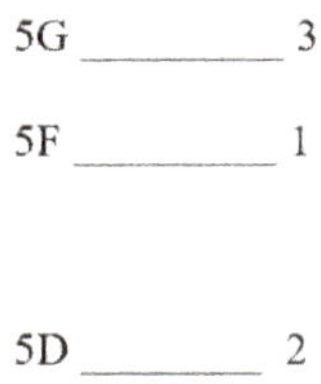

Figure A.4. The same as figure A.2 but with a different numeration of the energy levels.

For comparison with the above situation, we consider below the same system under a uniform electric field **F**. We slightly change the numeration of these energy levels, as shown in figure A.4.

The energy is counted from level one. Its eigenvalues are denoted by w. We introduce the following notations:

$$e = -E_2 > 0, \quad u = w/e, \quad a = z_{12}F/e, \quad b = z_{13}F/e, \quad c = E_3/e < 1. \tag{A.12}$$

The eigenvalues are the roots of the cubic equation:

$$u^3 + (1 - c)u^2 - (a^2 + b^2 + c)u - (b^2 - ca^2) = 0. \tag{A.13}$$

The strong-field approximation corresponds to the case in which

$$(a^2 + b^2)^{1/2} \gg 1. \tag{A.14}$$

Under this condition, the roots of the cubic equation (A.13) are as follows:

$$u_1 = -(b^2 - ca^2)/(a^2 + b^2), \tag{A.15}$$

$$u_2 = (a^2 + b^2)^{1/2} - d, \tag{A.16}$$

$$u_3 = -(a^2 + b^2)^{1/2} - d, \tag{A.17}$$

where

$$d = (1 - c - u_1)/2. \tag{A.18}$$

It should be noted that in the above results, obtained in the strong-field approximation, the quantities u_1 and d do not depend on the field (they are constants), while u_2 and u_3 linearly depend on the field. For comparison, figure A.5 presents the *exact* results for the energies versus the scaled field b for $a = (9/5)^{1/2}b$ and $c = 1/8$ (as is the case for the levels from figure A.4). It can be seen that as the scaled field b increases, the exact results approach the above asymptotic results obtained in the strong-field approximation.

It is instructive to represent the exact results for the energies versus b^2, that is, versus the scaled *energy density* of the field. This is shown in figure A.6. It can seen

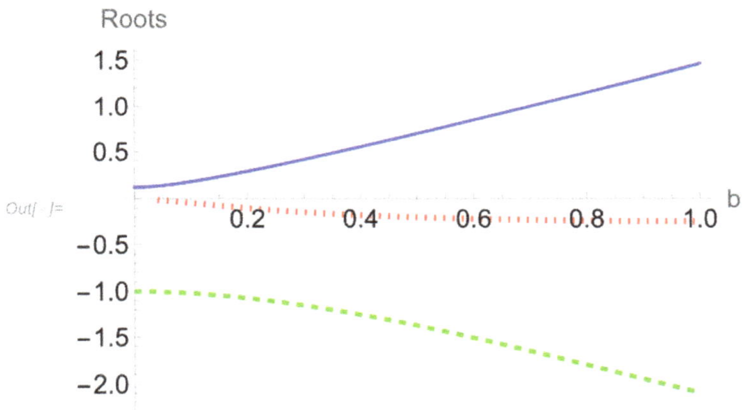

Figure A.5. The *exact* results for the energies versus the scaled field b for $a = (9/5)^{1/2}b$ and $c = 1/8$ (as is the case for the levels from figure A.4).

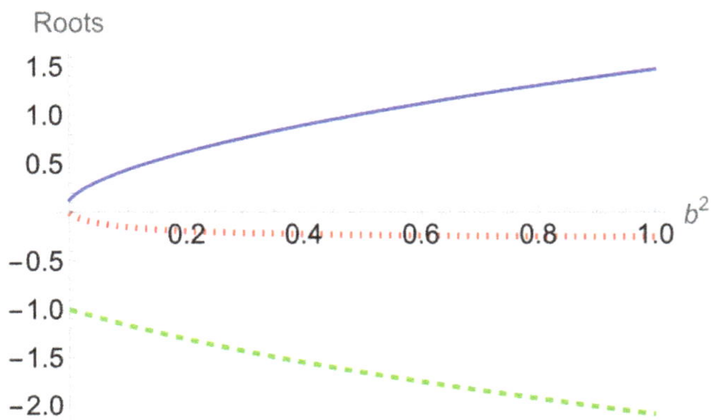

Figure A.6. The same results as those of figure A.5, but versus b^2, that is, versus the scaled *energy density* of the field.

that the dependence of the energies is not as highly nonlinear as in the case of the quasienergies presented in figure A.3.

Bibliography

[1] Baranger M and Mozer B 1961 *Phys. Rev.* **123** 25
[2] Kunze H J and Griem H R 1968 *Phys. Rev. Lett.* **21** 1048
[3] Cooper W S and Ringler H 1969 *Phys. Rev.* **179** 226
[4] Oks E 1995 *Plasma Spectroscopy: The Influence of Microwave and Laser Fields* (Berlin: Springer)

Appendix B

Examples of the use of spectral line radiation to perform diagnostics of oscillatory electric fields in laser-produced plasmas

The authors of [1] were the first to use the spectral line radiation (specifically, in the x-ray range) to perform diagnostics of the oscillatory electric fields in laser-produced plasmas. The experiment was conducted at the nanosecond Nd:glass laser facility at LULI (France). The laser intensity at the target was relatively moderate: 2×10^{14} W cm^{-2}. The target was structured: it consisted of a (20–60) μm thick powdered aluminum carbide Al_4C_3 strip placed between substrates made of plastic or magnesium. The plasma parameters (estimated by hydrodynamic simulations) at the surface of the target were an electron density of $N_e \sim 5 \times 10^{22}$ cm^{-3} and a temperature of $T \sim 300$ eV.

The Vertical Johann Spectrometer produced two sets of spatially resolved spectra simultaneously. The two sets were symmetrically located in relation to the unperturbed wavelength of the spectral line: the right set and the left set. This feature very significantly enhanced the possibility of identifying tiny details (Langmuir 'dips') within the experimental line profiles.

Figure B.1 shows the experimental profile of the Lyman-gamma line of Al XIII. In this figure, the right set and the left set are superimposed. This demonstrates the reproducibility of the Langmuir dip position (marked b_2) in the blue part of the experimental profile. The Langmuir-wave-induced 'bump-dip-bump' structures, being imposed on the inclined unperturbed profile, can create a secondary minimum (in addition to the primary minimum, whose location is controlled by the electron density) of no practical importance. The secondary minimum is marked by the arrow in the blue part of the experimental profile. The structures marked in the red part of the experimental profiles are the charge-exchange-induced dips—irrelevant to the phenomenon of Langmuir-wave-induced dips.

doi:10.1088/978-0-7503-5552-0ch10

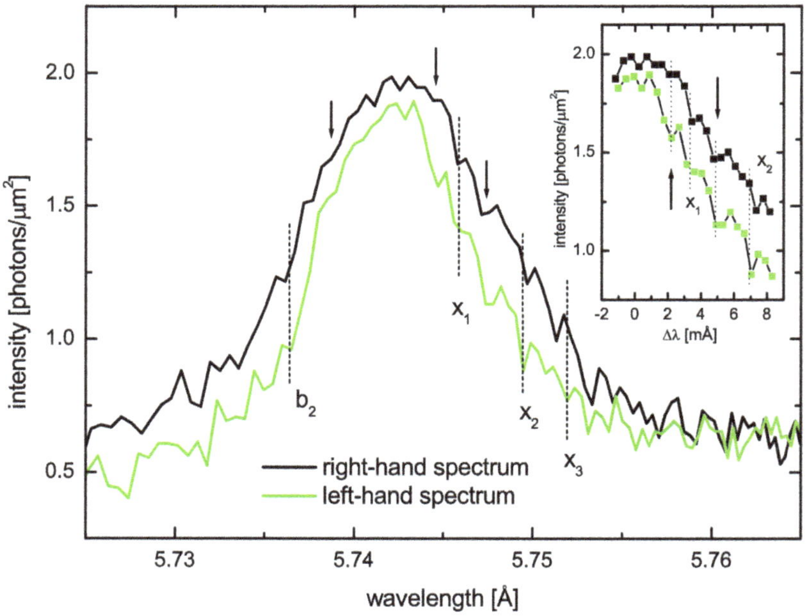

Figure B.1. The experimental profile of the Lyman-gamma line of Al XIII. In this figure, the right set and the left set are superimposed. This demonstrates the reproducibility of the Langmuir dip position (marked b_2) in the blue part of the experimental profile—see the text above the figure for more details. Reprinted from [1], Copyright (2006), with permission from Elsevier.

From the experimental position of the Langmuir dip, the electron density was determined to be 2.2×10^{22} cm^{-3}. From the width of the experimental Langmuir dip, the amplitude of the Langmuir wave was estimated to be 0.1 GV cm^{-1}.

The authors of [2] analyzed the experimental profiles of the Al XII beta line emitted from the aluminum plasma. The experiment was conducted at the Friedrich Schiller University Jena laser system, JETi.

The plasma was produced by a picosecond laser beam and then exposed to the beam of another picosecond laser. The second laser was turned on after the pulse of the first laser had terminated.

The first laser beam had an intensity of 5×10^{15} W cm^{-2}. The second laser beam had an intensity of 1.2×10^{16} W cm^{-2}. Both beams overlapped in space and were synchronized in time with a precision exceeding 1 ps.

Figure B.2 demonstrates the reproducibility of the structures in the experimental profiles of the Al XII beta line.

Simulations employing the Floquet–Liouville formalism [3] were conducted for electron densities of 2.5×10^{20} cm^{-3} $\leqslant N_e \leqslant 1.5 \times 10^{21}$ cm^{-3} at a temperature of $T_e = 150$ eV for a variety of field strengths of the quasimonochromatic electric field: $E = 0$, 0.1, 0.3, and 0.6 in units of $E_0 = E_0 = 5.146$ GV cm^{-1}, the latter being the atomic unit of the electric field. The results are presented in figure B.3 (for the 'parallel' profiles, i.e. those polarized along the oscillatory electric field) and in

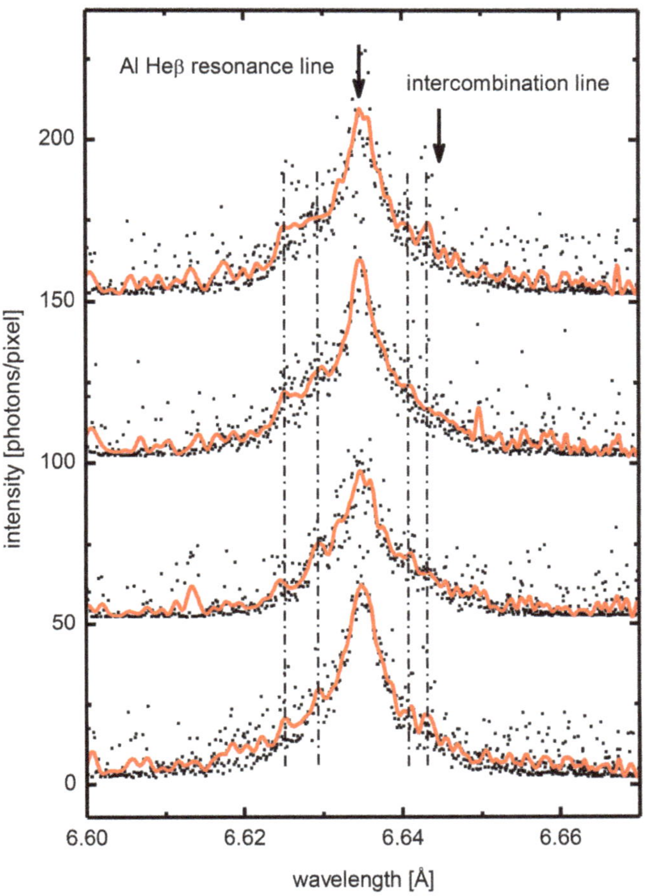

Figure B.2. A demonstration of the reproducibility of the structures in the experimental profiles of the Al XII beta line. Reproduced from [2].

figure B.4 (for the 'perpendicular' profiles, i.e. those polarized perpendicular to the oscillatory electric field).

Figure B.5 displays the outcome of kinetic PIC simulations of the distribution (in space) of the transverse averaged oscillatory electric field inside the plasma for $N_e = 0.1\ N_c$, $N_e = 0.2\ N_c$, and $N_e = 0.3\ N_c$. Here, $N_c = 1.7 \times 10^{21}\ \text{cm}^{-3}$ is the critical density. The nonlinear enhancement of the incident laser field occurs at the surface of the critical density.

Figure B.6 shows a comparison of the experimental profiles with the profiles simulated for the following four values of the initial electron density: 0.05 N_c, 0.1 N_c, 0.2 N_c, and 0.35 N_c.

Figure B.7 presents a comparison between the experimental profile and a profile simulated for an average electron density of $\bar{N}_e = 0.2\ N_c = 3.3 \times 10^{20}\ \text{cm}^{-3}$ and a time interval $\Delta T_{\text{unpert}} = 8.9$ ps. The comparison shows good agreement in the central part of the profile as well as in the first two satellites in the red and blue parts of the

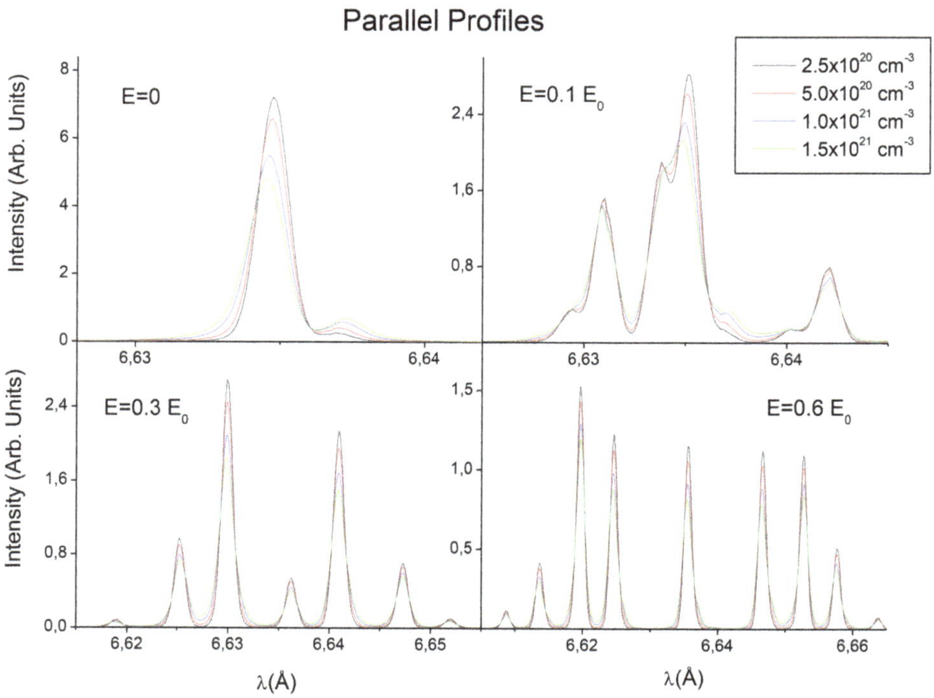

Figure B.3. Simulated profiles of the Al XII beta line at $T_e = 150$ eV for the indicated range of electron density and the indicated amplitude range E of the oscillatory electric field, the latter being in units of $E_0 = 5.146$ GV cm^{-1}. The profiles are polarized along the oscillatory electric field. Reproduced from [2].

profile. More distant satellites are in the far wings: there, the experimental profile blends with the noise. An additional maximum marked by the arrow corresponds to the dipole forbidden transition 1s3s $^1S_0$ — 1s^2 $^1S_0$ caused by the quadrupole interaction with the plasma ion field.

The authors of [4] analyzed the experimental hydrogenic spectral lines of Si XIV and of Al XIII produced during the interaction between relativistic laser radiation (at an intensity of ~10^{21} W cm^{-2}) and thin foils of silicon. As a result, the authors of [4] revealed the development of 'ionic sound' (i.e. the ion acoustic turbulence). This marked the discovery of ionic sound in in spectroscopic studies of laser-produced plasmas. Ionic sound developed at the surface of the critical density due to the parametric decay instability in the plasma. This experiment was conducted at the Vulcan petawatt laser facility located at the Rutherford Appleton Laboratory [5].

Parametric decay instability is a nonlinear process in plasmas. In this process, the electromagnetic wave is converted into an ionic sound wave and a Langmuir wave, both of which represent electrostatic plasma turbulence. The electrostatic plasma turbulence corresponds to the collective degrees of freedom in those plasmas.

The ion acoustic turbulence strongly influences transport phenomena in variety of plasmas. In particular, the resistivity of the plasma becomes anomalous: the primary cause of resistivity becomes the scattering of the conductivity electrons on the ion

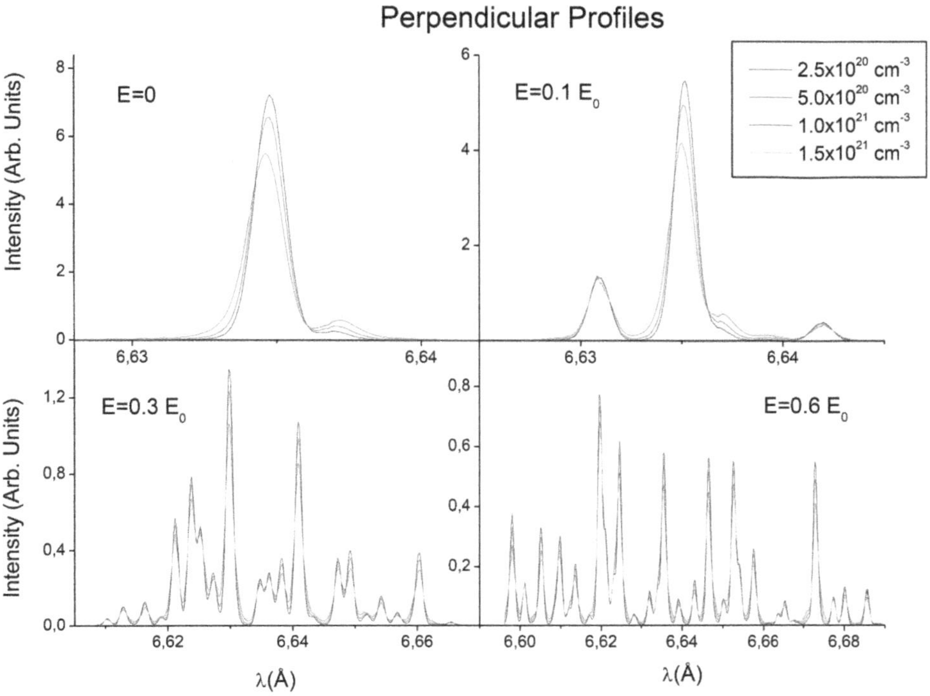

Figure B.4. The same profiles as those of figure B.3 but polarized perpendicular to the oscillatory electric field. Reproduced from [2].

acoustic wave—rather than the scattering on individual plasma ions (the latter being the cause of the 'normal' resistivity of plasmas).

Figure B.8 displays the experimental setup and the experimental profiles of the Ly-beta and Ly-gamma lines of Si XIV. In the insets, the locations of the Langmuir dips are indicated by vertical line segments. The distance between the experimental Langmuir dips is either $2\lambda_{pe}$ or $4\lambda_{pe}$, where

$$\lambda_{pe} = [\lambda_0^2/(2\pi c)]\omega_{pe}. \tag{B.1}$$

In equation (B.1), λ_0 is the unperturbed wavelength of the corresponding line and ω_{pe} is the plasma electron frequency:

$$\omega_{pe} = (4\pi e^2 N_e/m_e)^{1/2}. \tag{B.2}$$

Figure B.9 presents a comparison between the experimental and theoretical profiles of the Ly-beta and Ly-gamma lines of Si XIV. The theoretical profiles were calculated by allowing both for the effect of the ion acoustic turbulence and for the Langmuir dips.

Figure B.10 displays the experimental profile of the Ly-beta line of Al XIII obtained when the spectrometer was directed at the front surface of the laser-irradiated target—in contrast to the experimental profiles in figures B.8 and B.9,

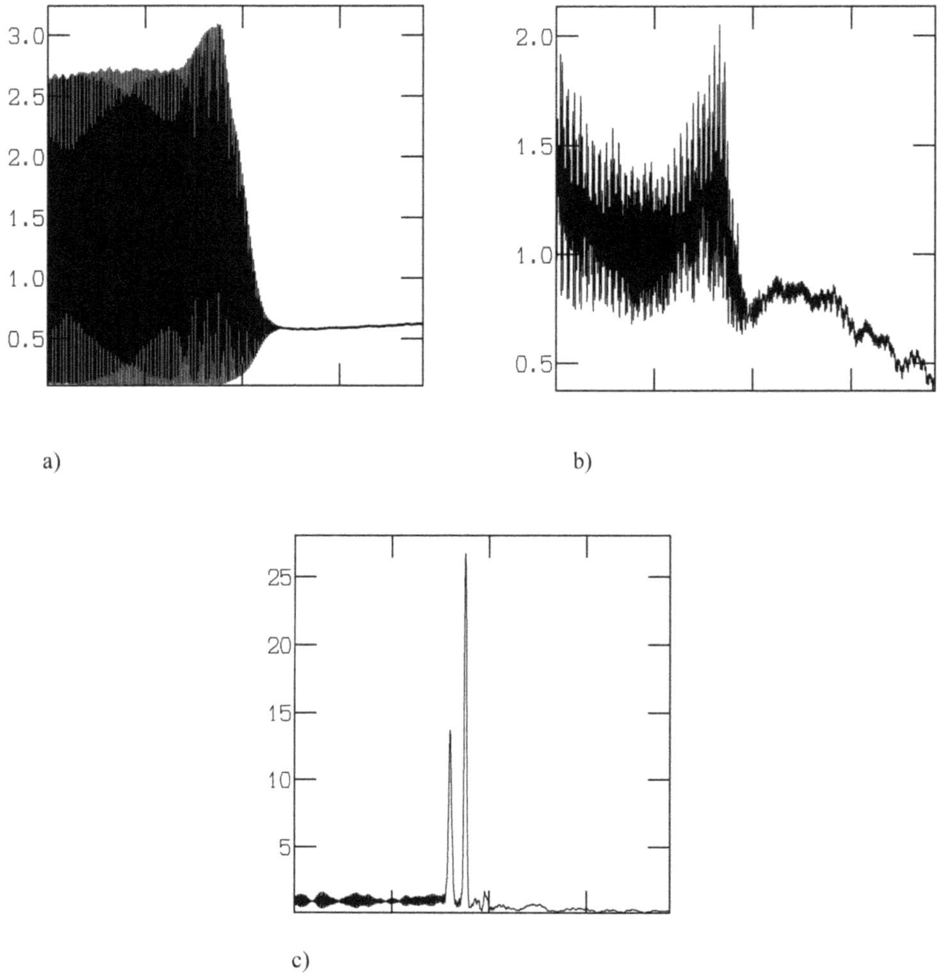

Figure B.5. The outcome of the kinetic PIC simulations of the distribution (in space) of the transverse averaged oscillatory electric field inside the plasma: (a) $N_e = 0.1N_c$; (b) $N_e = 0.2N_c$; (c) $N_e = 0.3N_c$. Here, $N_c = 1.7 \times 10^{21}$ cm^{-3} is the critical density. Reproduced from [2].

which were obtained by directing the spectrometer at the back surface of the laser-irradiated target. The target was aluminum foil coated by a layer of CH. The corresponding theoretical profile is also presented. The experimental spectrum exhibits two pairs of Langmuir dips: one pair at a distance of ±33.6 mÅ from the line center and the second pair at a distance of ±16.8 mÅ from the line center.

The authors of [6] presented an in-depth spectroscopic study of the simultaneous production of Langmuir waves and the ionic sound at the surface of the relativistic critical density. This experiment was also conducted at the Vulcan petawatt laser facility located at the Rutherford Appleton Laboratory [5]. The focus of the paper was the Langmuir dips in the experimental profiles of the Ly-beta line of Si XIV. The laser intensity at the target ranged from 1 to 3×10^{20} W cm^{-2} in different shots.

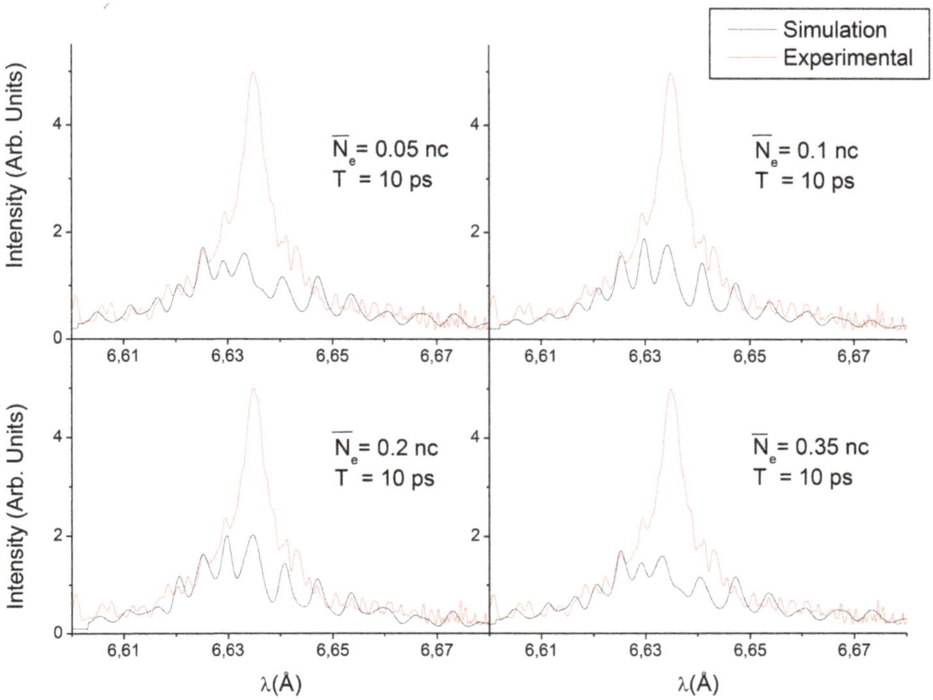

Figure B.6. A comparison between the experimental and simulated profiles for the following four values of the initial electron density: 0.05 N_c, 0.1 N_c, 0.2 N_c, and 0.35 N_c. Reproduced from [2]. © IOP Publishing Ltd. All rights reserved.

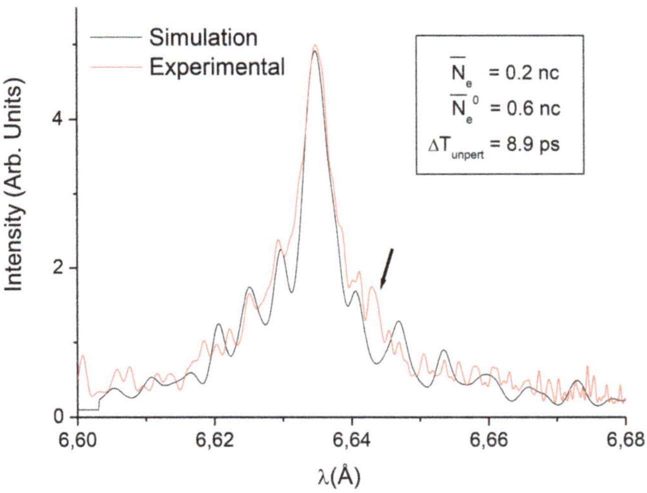

Figure B.7. A comparison between the experimental profile and a profile simulated for an average electron density of $\bar{N}_e = 0.2\ N_c = 3.3 \times 10^{20}\ \mathrm{cm}^{-3}$ and a time interval of $\Delta T_{\mathrm{unpert}} = 8.9$ ps. Reproduced from [2]. © IOP Publishing Ltd. All rights reserved.

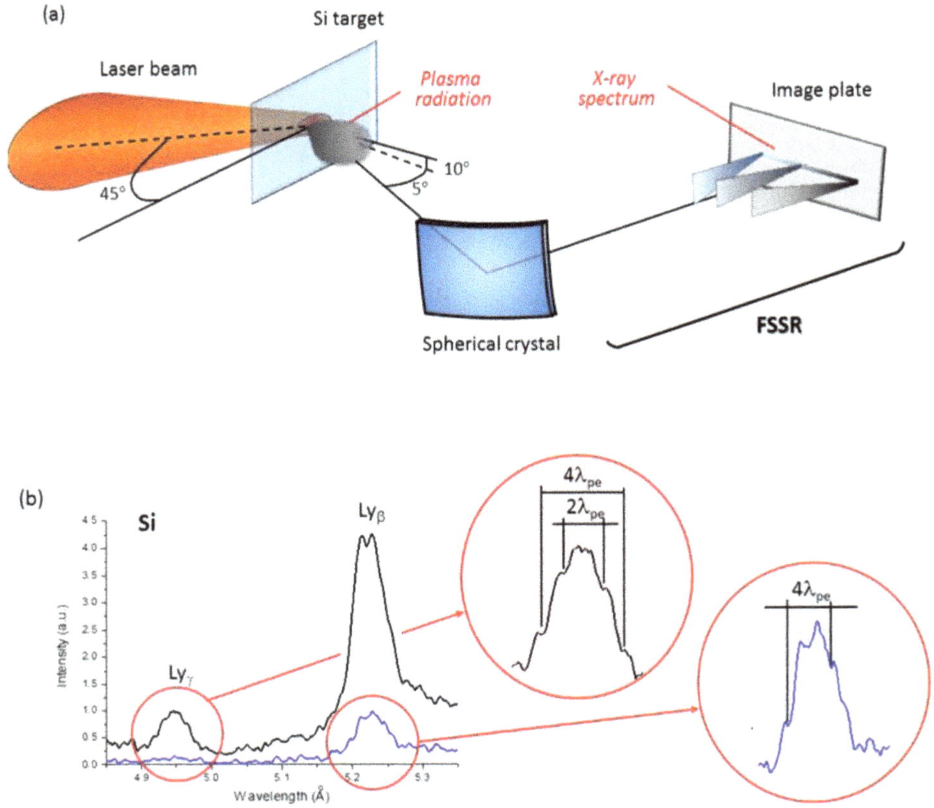

Figure B.8. (a) The experimental setup and (b) the experimental profiles of the Ly-beta and Ly-gamma lines of Si XIV. In the insets, the locations of the Langmuir dips are indicated by the vertical line segments. Reproduced from [4]. © The Optical Society.

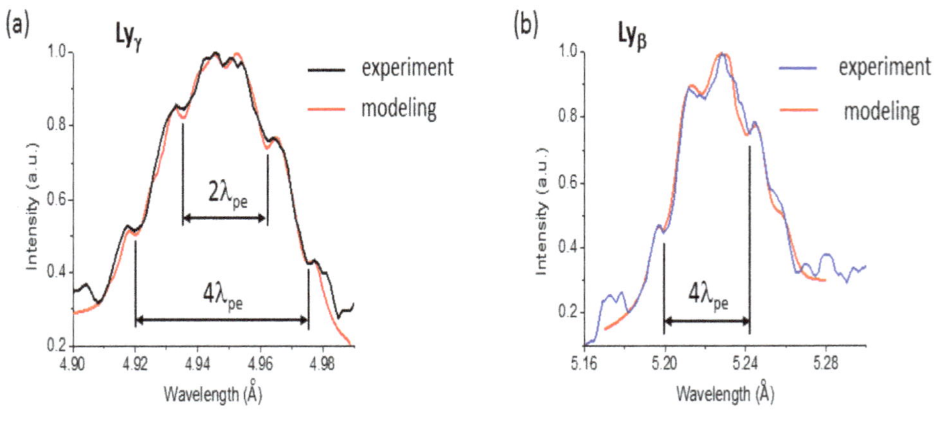

Figure B.9. A comparison between the experimental and theoretical profiles of the Ly-beta and Ly-gamma lines of Si XIV. The theoretical profiles were calculated by allowing both for the effect of the ion acoustic turbulence and for the Langmuir dips. The theoretical profiles are calculated for the electron density $N_e = 1.74 \times 10^{22}$ cm^{-3}. Reproduced from [4]. © The Optical Society.

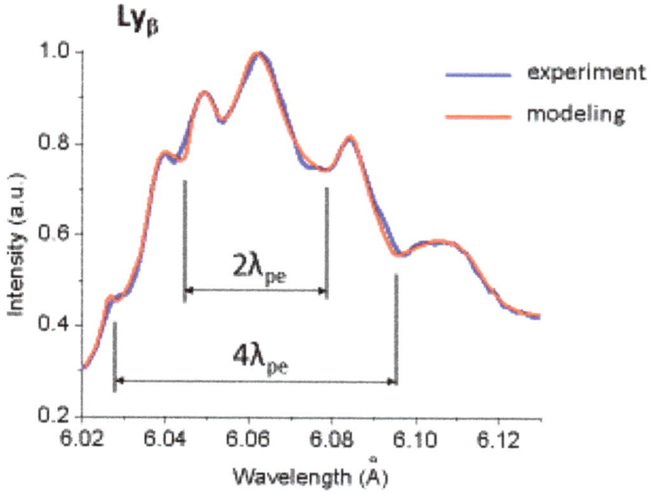

Figure B.10. The experimental profile of the Ly-beta line of Al XIII obtained by directing the spectrometer at the front surface of the laser-irradiated target—in contrast to the experimental profiles in figures B.8 and B.9, which were obtained by directing the spectrometer at the back surface of the laser-irradiated target. The target was aluminum foil coated by a layer of CH. The corresponding theoretical profile is also presented. Reproduced from [4]. © The Optical Society.

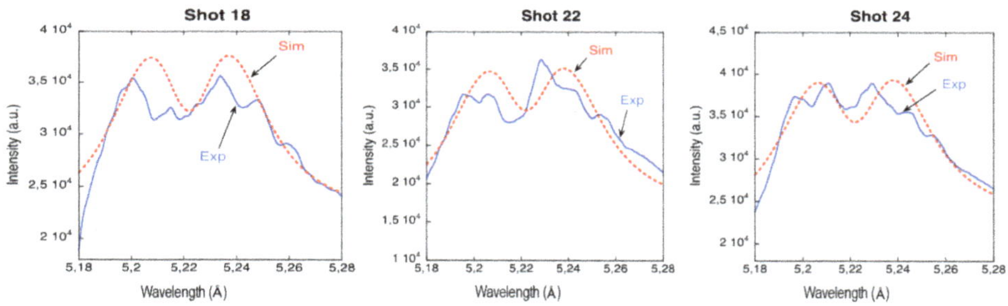

Figure B.11. Experimental profiles of the Si XIV Ly-beta line obtained in three shots. Also shown are the corresponding profiles simulated by the FLYCHK code. The best fit produced by this code yielded an unrealistically high electron density of $N_e = 6 \times 10^{23}$ cm^{-3}. Reproduced from [6]. © IOP Publishing Ltd. All rights reserved.

Figure B.11 displays experimental profiles obtained in three different shots. There are very distinct 'bump-dip-bump' structures in the experimental profiles. Actually, these are Langmuir superdips, that is, the superposition of two Langmuir dips at the same location in the profile. Also shown are theoretical profiles calculated using the FLYCHK code, which does not take into account the phenomenon of the Langmuir dips and the broadening due to ion acoustic turbulence. The best fit by this code yielded an unrealistically high electron density of $N_e = 6 \times 10^{23}$ cm^{-3}. This was one of the indications that ion acoustic turbulence had developed and that it has to be taken into account.

Figure B.12 shows the experimental profiles obtained in the shots marked A, B, and C. Also shown are the corresponding theoretical profiles calculated by more advanced code that took account of the phenomenon of the Langmuir dips and the broadening due to ion acoustic turbulence—in addition to all other broadening mechanisms. The deduced electron density was $N_e = 2.2 \times 10^{22}$ cm^{-3}. The root-mean-square strengths of the electric field of the ion acoustic turbulence were deduced to be 4.8 GV cm^{-1}, 4.4 GV cm^{-1}, and 4.9 GV cm^{-1} for shots A, B, and C, respectively.

Shot D was obtained at a significantly lower laser intensity at the target than shots A, B, and C. Therefore, in shot D the electron density was significantly smaller than in shots A, B, and C. The consequence was a significantly smaller damping of the Langmuir waves, so that they attained a considerably greater amplitude.

The deduced parameters for shot D were an electron density of $N_e = 6.6 \times 10^{21}$ cm^{-3}, a temperature of T = 550 eV, a root-mean-square electric field of the ion acoustic turbulence of $F_{t,rms}$ = 2.0 GV cm^{-1}, and a Langmuir wave amplitude of

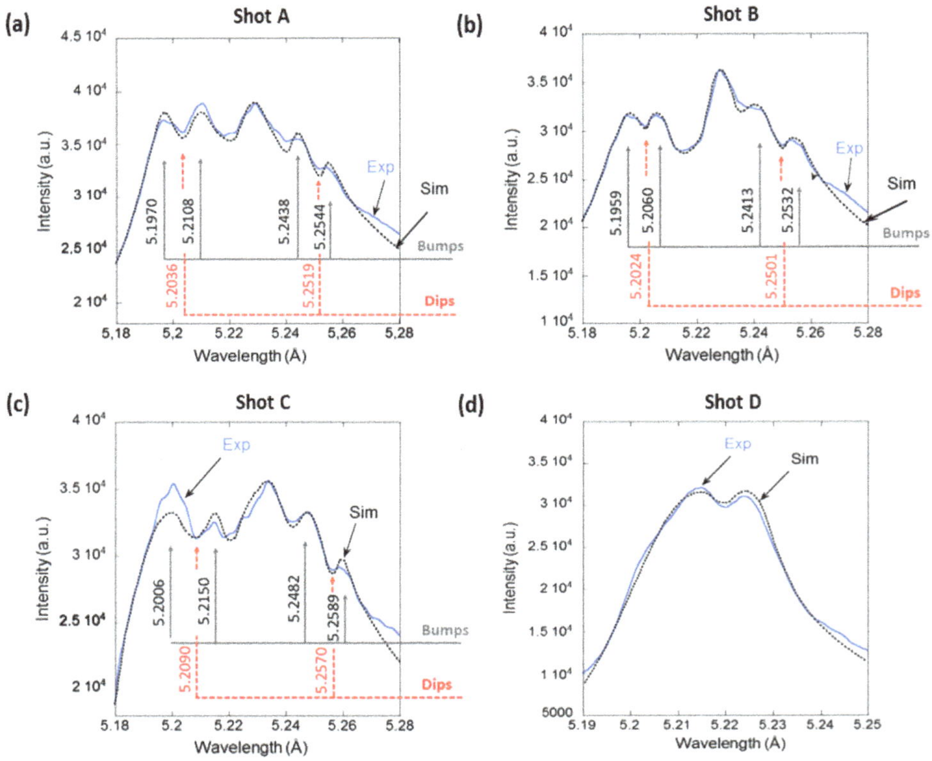

Figure B.12. Experimental profiles of the Si XIV Ly-beta line obtained in four different shots and their comparison with the profiles calculated by more advanced code that took into account the phenomenon of the Langmuir dips and the broadening due to ion acoustic turbulence (for shots A, B, and C)—in addition to all other broadening mechanisms. The deduced electron density was $N_e = 2.2 \times 10^{22}$ cm^{-3}. The root-mean-square strengths of the electric field of the ion acoustic turbulence were deduced to be 4.8 GV cm^{-1}, 4.4 GV cm^{-1}, and 4.9 GV cm^{-1} for shots A, B, and C, respectively. Reproduced from [6].

$E_0 = 2.0$ GV cm^{-1}. The theoretical and experimental profiles in shot D do not exhibit bump-dip-bump structures. This is because in the situation in which $E_0/F_{t,rms} > 0.5$, Langmuir dips cannot form, as explained in [7].

Bibliography

[1] Renner O, Dalimier E, Oks E, Krasniqi F, Dufour E, Schott R and Förster E 2006 *J. Quant. Spectrosc. Radiat. Transfer* **99** 439

[2] Sauvan P, Dalimier E, Oks E, Renner O, Weber S and Riconda C 2009 *J. Phys. B: At. Mol. Opt. Phys.* **42** 195001

[3] Sauvan P and Dalimier E 2009 *Phys. Rev.* E **79** 036405

[4] Oks E *et al* 2017 *Opt. Express* **25** 1958

[5] Danson C N *et al* 2005 *Laser Part. Beams* **23** 87

[6] Oks E *et al* 2017 *J. Phys. B: At. Mol. Opt. Phys.* **50** 245006

[7] Oks E 1995 *Plasma Spectroscopy: The Influence of Microwave and Laser Fields, (Springer Series on Atoms and Plasmas)* 9 (New York: Springer)

Appendix C

The validity of using the analytical method based on separating rapid and slow subsystems for a more accurate analysis of the Stark broadening of hydrogenlike spectral lines by plasma electrons

The frequency of the spectral line is the frequency of the revolution of the atomic electron around the nucleus:

$$\Omega = \frac{Z^2 U_H}{\hbar}\left(\frac{1}{n_b^2} - \frac{1}{n_a^2}\right), \tag{C.1}$$

where U_H is the ionization potential of hydrogen.

The typical frequency characterizing the motion of the perturbing electron around the radiating ion, leading to the Stark broadening of spectral lines, is the Weisskopf frequency (see, e.g. [1]):

$$\omega_{We} = \frac{v_T}{\rho_{We}} \sim \frac{Z\, m\, v_T^2}{(n_a^2 - n_b^2)\hbar} \sim \frac{Z\, T}{(n_a^2 - n_b^2)\hbar}. \tag{C.2}$$

Consequently, the ratio of the above frequencies is given by

$$\frac{\omega_{We}}{\Omega} \sim \left(\frac{T}{Z\, U_H}\right)\left[\frac{n_a^2 n_b^2}{(n_a^2 - n_b^2)^2}\right]. \tag{C.3}$$

In the situation in which $n_a \gg n_b$, this ratio simplifies as follows:

$$\frac{\omega_{We}}{\Omega} \sim \left(\frac{T}{Z\, n_a^2 U_H}\right) \ll 1 \tag{C.4}$$

if

$$T(\mathrm{eV}) \ll (13.6\mathrm{eV}) Z n_{\mathrm{a}}^{2}. \tag{C.5}$$

Thus, under condition (C.5), it is legitimate to apply the method of separating rapid and slow subsystems to the problem under consideration.

For instance, for the case of ionized helium ($Z = 2$), the above validity condition (C.5) becomes

$$T\ (\mathrm{eV}) \ll (27.2\mathrm{eV})\ n_{\mathrm{a}}^{2}. \tag{C.6}$$

The inequality (C.6) is met for a sufficiently large range of temperatures at which the spectral line radiation of ionized helium is observed in various plasmas.

Bibliography

[1] Oks E 2006 *Stark Broadening of Hydrogen and Hydrogenlike Spectral Lines in Plasmas: The Physical Insight* (Oxford: Alpha Science International)

Appendix D

A brief outline of the impact approximation in the conventional theory of the Stark broadening of spectral lines in plasmas

For the dynamical Stark broadening of spectral lines by plasma electrons to be efficient, it is required that the formally calculated number ν_{We} of perturbing electrons located within the sphere of the electron Weisskopf radius ρ_{We} should be much smaller than unity (see, e.g. [1]):

$$\nu_{We} = 4\pi N_e \rho_{\mathrm{We}}{}^3/3 \ll 1. \tag{D.1}$$

In equation (D.1), $\rho_{\mathrm{We}} \sim n^2\, \hbar/(m_e v_{\mathrm{Te}})$, where n is the principal quantum number and v_{Te} is the mean thermal velocity of the perturbing electrons in a plasma. When inequality (D.1) is satisfied, the instantaneous Stark splitting in the electron microfield is much smaller than the typical frequency of the variation in the electron microfield $\Omega_e \sim v_{\mathrm{Te}}/\rho_{\mathrm{We}}$ (for the overwhelming majority of perturbing electrons). Physically, the electron Weisskopf radius is associated with values of the impact parameters $\rho \sim \rho_{\mathrm{We}}$, which provide the primary contribution to the electron-induced dynamical Stark broadening of spectral lines in plasmas [2, 3].

The essence of the electron-induced dynamical Stark broadening is as follows: *virtual transitions* within the lower (n') and upper (n) multiplets occur in the process of the radiative transition $n \leftrightarrow n'$. Their principal effect is to shorten the lifetimes of the states n' and/or n; this causes the broadening of the line.

The assumption that virtual transitions take place mostly within the upper and lower multiplets is called the *no-quenching approximation*. Within this approximation, virtual transitions between states that differ by the principal quantum numbers are not taken into account. This approximation makes it possible to use the concept of the *line space*. The latter is the direct product of the Hilbert space spanned on the basis vectors of the n-subspace with the Hilbert space spanned on the (complex--conjugated) basis vectors of the n'-subspace.

doi:10.1088/978-0-7503-5552-0ch12

Not only the impact formalism (introduced in [2] and further developed in [4]), but also the primary properties of the unified theory of the electron-induced dynamical Stark broadening [5] can stem from the same succession of mathematical operations, as shown in [6]. The main difference between these two formalisms is as follows.

The impact formalism treats all collisions as being completed. In contrast, the unified theory of the dynamical Stark broadening takes into account the fact that that not all collisions are completed. Another difference between the two formalisms is that within the unified theory, the calculated Stark profile of the spectra line contains a transition to the quasistatic shape in the wings; however, the unified formalism does not always produce this transition in the correct way.

The crucial feature of the impact formalism is that it introduces a *coarse-grained timescale* Δt defined through the following condition:

$$\rho/v_e \ll \Delta t \ll \left[\max\left(\gamma, \Delta\omega, \omega_{pe}\right)\right]^{-1}. \qquad \text{(D.2)}$$

In equation (D.2), $\omega_{pe} = (4\pi N_e e^2/m_e)^{1/2}$ is the plasma electron frequency, $\Delta\omega$ is the detuning from the unperturbed frequency of the spectral line, and γ is the inverse of the lifetime of the radiator (the latter being of the order of the impact Stark width of the line).

The physical meaning of this coarse-grained timescale is as follows. The focal point is the radiator evolution in time during the intervals $\sim\Delta t$ defined by equation (D.2), while the details of the time evolution during the intervals $\sim\rho/v_e$ (i.e. during intervals of the order of the typical time of the individual collision) are of no interest.

The validity limits of the impact formalism are controlled by whether or not it is possible to introduce the coarse-grained timescale defined by the inequality (D.2). It is worth mentioning that in the unified theory, in the left part of the condition (D.2), the strong inequality $\rho/v_e \ll \Delta t$ is lessened to $\rho/v_e \sim \Delta t$.

The operator of the dynamical electron broadening Φ_{ab} is defined as follows:

$$\Phi_{ab} = [U_a(t, t + \Delta t)U_b^*(t, t + \Delta t) - 1]/\Delta t. \qquad \text{(D.3)}$$

In equation (D.3), U_a and U_b are the operators of the evolution in time for the upper and lower subspaces in the line space, respectively. If $\Delta t \gg \rho/v_e$ (i.e. under the impact approximation), these operators are replaced by the respective scattering matrices S_a and S_b. The utilization of the two scattering matrices (rather than one) is associated with the concept of the line space.

A final note: for any hydrogenic spectral line of any spectral series, dynamical Stark broadening takes place in the overlapping situation—that is, there are overlaps of the Stark components of the line. This is the case even for the Lyman-alpha line (the simplest line) of hydrogenic atoms/ions—see [7].

Further details on the impact approximation and its comparison with the unified theory can be found in [8–11]. Also, [6] is quite illuminating on this issue.

Bibliography

[1] Lisitsa V S 1977 *Sov. Phys. Usp.* **122** 603

[2] Baranger M 1958 *Phys. Rev.* **111** 481

[3] Baranger M 1958 *Phys. Rev.* **112** 1958 855

[4] Kolb A C and Griem H R 1958 *Phys. Rev.* **111** 514
[5] Vidal C R, Cooper J and Smith E W 1970 *J. Quant. Spectrosc. Radiat. Transfer* **10** 1011
Vidal C R, Cooper J and Smith E W 1970 *J. Quant. Spectrosc. Radiat. Transfer* **11** 263
[6] Sahal-Brechot S 1969 *Astron. Astrophys.* **1** 91
[7] Strekalov M L and Burshtein A I 1972 *Sov. Phys. JETP* **34** 53 http://jetp.ras.ru/cgi-bin/e/index/e/34/1/p53?a=list
[8] Griem H R 1964 *Plasma Spectroscopy* (New York: McGraw-Hill)
[9] Griem H R 1974 *Spectral Line Broadening by Plasmas* (Cambridge, MA: Academic) https://shop.elsevier.com/books/spectral-line-broadening-by-plasmas/griem/978-0-12-302850-1
[10] Oks E 2006 *Stark Broadening of Hydrogen and Hydrogenlike Spectral Lines in Plasmas: The Physical Insight* (Oxford: Alpha Science International)
[11] Oks E 2017 *Diagnostics of Laboratory and Astrophysical Plasmas Using Spectral Lines of One-, Two-, and Three-Electron Systems* (Hackensack, NJ: World Scientific)

www.ingramcontent.com/pod-product-compliance
Ingram Content Group UK Ltd.
Pitfield, Milton Keynes, MK11 3LW, UK
UKHW051851100926
8556IPUK00005B/568

9 780750 355506